Stahlhelm

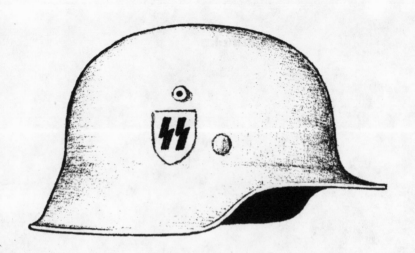

STAHLHELM

Evolution of the German Steel Helmet

Revised and Expanded Edition

Floyd R. Tubbs

with the assistance of

Robert W. Clawson

THE KENT STATE UNIVERSITY PRESS

Kent, Ohio, & London

©1971, 2000 by Floyd R. Tubbs
All rights reserved
Library of Congress Catalog Card Number 00-036875
ISBN 0-87338-677-9
Manufactured in the United States of America

Revised and expanded edition.

06 05 04 03 02 01 00   5 4 3 2 1

Library of Congress Cataloging-in-Publication Data

Tubbs, Floyd R., 1928–
Stahlhelm: evolution of the German steel helmet/by Floyd R.
Tubbs, with the assistance of Robert W. Clawson—Rev. and
expanded ed.
p. cm.
ISBN 0-87338-677-9 (pbk.: alk. paper) ∞
1. Helmets—Germany. 2. Germany—Armed Forces—
Equipment. I. Clawson, Robert W. II. Title

UC505.G3 T83 2000
623.4'4—dc21   00-036875

British Library Cataloging-in-Publication data are available.

Contents

Preface

IN THE LONG HISTORY of warfare and armies, the concept of military clothing in the form of a prescribed national headgear is of relatively recent origin. The twentieth-century steel helmet is one of the most important pieces of equipment in the armed forces. Its value was recognized early by Germany and is still used extensively.

German history is largely responsible for the development of the country's steel helmet. To understand the reasons for this development, I have included in this volume a summary of both political and military history. My intent is to link together the time involved before, during, and after the two world wars. This will enable readers better to acquaint themselves with the eras in which these particular helmets were used.

Each helmet has a historical tale to tell. Helmets also distinguish members of a certain armed service from civilians, as well as superiors from subordinates and friends from foes. They foster a pride of service. They provide a means by which the various units might be readily identified in action or on the march. Above all is the most valuable aspect of helmets: They save lives.

Today helmets are collectors' items. Headgear is one of the largest interests among militaria collectors and followers of military research, and the subject of the modern steel helmet is constantly expanding. As with all collectible items, there is always a longing for worthwhile information. This prompted my search for additional helmet literature. A benefit of compiling the data found in this text was discovering answers to two questions: "Why?" and "When?" There will always be a search for purposes, reasons. It is difficult to imagine the abundance of knowledge advancing through the years when the explanation of purpose is known to so few and is soon forgotten. Cumulative knowledge can only be built through the sharing of information that focuses on "Why" and "When."

This revised and expanded edition reflects data that has been made available since the original publication of *Stahlhelm* in 1971 as well as additional original research. Dr. Robert W. Clawson, emeritus professor of European military studies at Kent State University, assisted me in the preparation of this revised volume as a fellow collector and scholar.

I hope that the new edition of *Stahlhelm* will promote additional clarification on the subject of the modern steel helmet.

Acknowledgments

A BOOK OF THIS SCOPE cannot be the work of only one person. Exploring a field such as this means relying on many other specialists. I wish to acknowledge my deep indebtedness to the many people who so willingly contributed to the compiling of this book. Their responses to my seemingly endless requests for further data and clarification proved invaluable.

I wish to extend my particular appreciation to the following persons for their kind assistance and dedication to the production of this book. Special thanks go to Dennis Lewis: commercial artist and collector of German headgear. His hobby proved a valuable asset in producing the unique illustrations featured throughout this text. To Robert Oswald, free-lance writer and historical lecturer, I am grateful for his invaluable contribution. Thanks to Russel Jardine, an advanced helmet collector and known throughout the world by those concerned with military helmets. We are quite fortunate to have had the privilege of photographing selections in his vast helmet collection. And David Powers, widely known in helmet collecting circles for his encyclopedic knowledge of the field, enthusiastically, and generously, offered his expertise in the revision and expansion of the original edition.

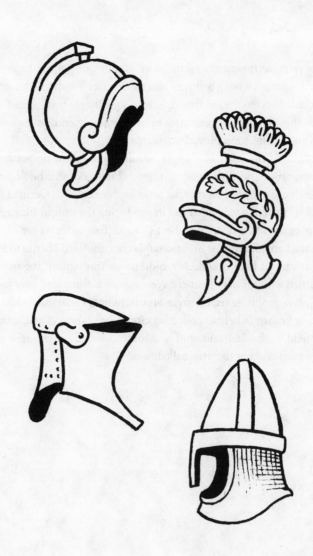

Author's Note

THERE ARE TWO CLASSIFICATIONS of combat helmets: armor and modern helmets.

From earliest recorded battles, the helmet—along with the shield, chain mail, and the suit of armor—has been used to help bring the warrior safely home. Today, with few exceptions, the helmet remains a universally recognized symbol of the warrior.

There was a period in which helmets were separated into two groups. Steel helmets were abandoned for several centuries, and armies were equipped with soft hats for battle. This remained so for a long time, and history does not adequately account for what seems like a paradox. The metal helmet was revived at the beginning of the twentieth century during the era of the two World Wars, or what came to be referred to as total war. Since then all helmets have become known as "modern helmets" and are not in the same classification as armor used prior to the twentieth century.

This book does not explore armor as the unique and esoteric field is covered by many excellent published studies. Where those works cover the first era of military helmets, *Stahlhelm* begins with the creation of the modern helmet.

THE PRESENT WAR IS reviving old methods. The steel fort has been discredited and the earthwork justified; the strength and direction of the wind are important in connection with aeroplanes and gas attacks; hand grenades and bombs have assumed importance; and the question of armor for the soldier has come up.

The French are about to supply a metal helmet to soldiers. Trench warfare has led to a preponderance of head wounds. The metal helmet has by practical test proved efficacious in preventing or lessening the severity of these wounds. The chief objection to the use of armor . . . loss of mobility . . . disappears in trench warfare.

— "Armour for Battle," *Times of London,* July 23, 1915

BESIDES THE SMALL PORTABLE shields which the various nations have adopted before the war for the protection of sappers and of infantry in the trenches, the French had experimented on shields to protect the body in front and on the sides as well as the legs. Dr. Devraigne of the French Army published an article in the *Lancet* setting forth the result of experiments with the metal helmets. In 42 cases of head wounds where the helmets were not used there were 23 serious fractures of the skull. In 13 cases where men had used the metal helmet, there were eight cerebral hemorrhages and five superficial wounds, with no deaths. The use of the metal helmet will probably become general.

— "Helmets and Shields for Protection from Projectiles,"
Revista de Artilharia, July 1915

An Old Helmet Has an Interesting Story to Tell

THE GERMAN ARMY LOST two World Wars in a period of twenty-seven years. Ironically, Germany may have gained more by losing than by winning. In the United States there is the maxim: "It matters not whether you win or lose, but how you play the game." In hindsight, it seems the German army "played the game" well and often with stunning brutality. Even the Allied armies gave begrudging appreciation to the enemy who refused to quit, even in the face of overwhelming superiority in numbers during both wars.

Millions of soldiers fought in World War I and World War II, and out of all of them the German soldier emerged as the ideal twentieth-century warrior: brave, loyal, efficient, ruthless. There are some who feel that the defeat of the German army in both wars was a matter of quantity defeating quality. But, in truth, a soldier can be found in every army in the world who can match the best German soldier.

While every army in existence today can point with pride to a number of spectacular victories, the German soldier is still the world's popular choice as the ideal soldier. For this honor, the German soldier must give proper credit to the man who designed his wardrobe. Allied soldiers in both World Wars cast envious eyes on the German soldiers' uniform. Psychologically, it is believed that the uniform worn by the German made him feel like a warrior. At the outbreak of WWII, those in the German army were the highest paid in the world. As the best-dressed and best-paid soldier, along with a given cause for vengeance, the German soldier of WWII was a chilling example of good soldiering. His confidence and esprit de corps carried him to the very brink of an impossible military accomplishment.

In both wars, the most distinctive feature of the Germany army uniform was the item that has come to symbolize German militarism in even the remotest corners of the world—the helmet, the Stahlhelm.

Two vintage items of World War I that the German military staff felt adaptable for use in the military techniques of World War II were the gun and the helmet. The gun, developed at the end of WWI, was inspected by the conquering Allied armies and was thought to be so inferior that it was ignored. In WWII it became known as the 88 mm, a mobile cannon that struck fear into the hearts of men who had to face it. In the same light, German military leaders felt the helmet needed no basic improvement. Those decisions, to revive the 88 mm and to continue with the m-16 and m-17/18 helmets and to modify them, proved to be master strokes.

At times during WWI, German troops moved around in a surrounding mist, only a soldier's helmet showing above the fog. An observer, seeing nothing but helmets and rifle barrels, remarked, "Damned if it doesn't seem like the helmets are going to war by themselves."

In May 1940 a citizen in a town in Holland pulled the corner of his curtain back and saw standing in the street nearby a soldier wearing a German helmet. The Dutchman was paralyzed with fear because the German helmet bore a terrible significance to him.

In a Hungarian town a low-ranked government official noticed that when the first German helmet appeared in town, the Communists disappeared. With mixed emotions, he connected the odd-shaped helmets of the German army with a symbol of anticommunism. Unfortunately for him, it was not a Hungarian helmet. "Too bad it's German," he thought. "I would like to put one on and go calling on some of those Socialist troublemakers."

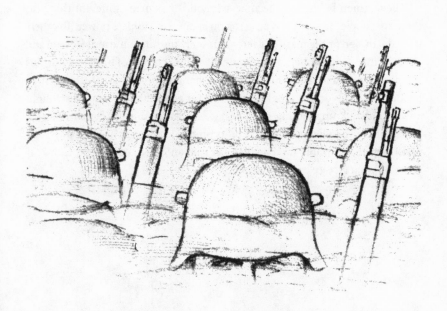

A Russian veteran of WWI watched Germans move through his village. He was one of millions of Russians who had surrendered to the Germans in 1915 and who had spent the war in a German POW camp. Fighting for the Tsar had not been his idea of a cause, and some had even looked upon the Germans as liberators at the time. But here he was once again looking at the familiar German helmet and wondered if he was dreaming. Perhaps his life was better under Stalin than it had been under Tsar Nicholas? But when he needed the Germans in 1919, they could not come, and now that he did not want them, here they were. Same helmets . . . same men?

In London, an ex-sergeant of the British army sat in a cinema watching a newsreel of the Germans invading Holland and cursed at the relative ease of their victory. Except for the greater mechanization, they were identical to the WWI troops. Two young men sitting beside him were engaged in conversation. One of the boys said, "Looks like we're going to be in it again. Only this time we won't take so long in doing the Germans in."

The ex-sergeant laughed out loud. "That's what I thought a few years ago, kid," he said. "You'll learn to respect the German, too, because he's a damned good soldier. And the back of his neck don't get wet."

To which the boy responded, "What do you mean the back of his neck doesn't get wet?"

"It's their helmet, kid, see? It covers his ears and the back of his neck and not just from the bullets. When it rains, the water don't run down the back of his neck, under his collar. Very uncomfortable feeling, that."

3

The German helmet has become symbolic of as many different things as there are people. To some it is an arrogant symbol of ruthless power. To others it is a death sign or the mark of a ferocious bully pushing innocent women and children into gas chambers. And to some it signifies an era of brief but glorious revenge on Germany's enemies. But the general impression that the German helmet arouses is one of toughness. Its wearer was a faceless man who fought a war in a cold, unrelenting manner. He was feared, but he feared nothing. Thus, the German helmet, because of its unique shape, and because it was the first thing recognizable on a soldier, became an object of visceral fear. It is amazing what a piece of inert metal can do!

ONE

The Summer of 1914,
Europe

THOUGH HISTORIANS DEFINE "peace time" as a period between wars, in truth, armed conflict between two or more major powers is labeled as "war." Therefore, in a sense, in the summer of 1914, before World War I began officially, the world was at war.

Japan had beaten mighty Russia to its knees just nine years before. It then began casting a predatory eye on the Pacific possessions of Portugal, Britain, France, Germany, the Netherlands, and the United States. Britain was physically extended to the limit and was engaged in conflicts in Ireland, Africa, and India. The French, with possessions in Africa, were fighting there and in Indochina and were suspicious of Britain's aims in those areas. Spain and Portugal were weak and seemed resigned to the eventual loss of their colonies.

Russia, rebuffed so harshly in the east by Japan, had turned its attention to the West and tried to bring the Serbs and other southern Slavs into its clutches. The Austro-Hungarian Empire, bordering on Russia, held millions of Slavs in its domain. Russia wooed the tiny nation of Serbia, and together they began a campaign of underground activities in Austria-Hungary to agitate the Serbs and other Slavs.

Turkey sat astride the Dardanelles, aware that Russia dreamed of a seaport on the Mediterranean in that exact spot. Bad feelings prevailed between Turkey and the Greeks, and the Ottoman Empire was made up of many nationalities anxious to break free, particularly the Armenians and Kurds.

Austria-Hungary was an empire that was comprised of eight nations, seventeen countries, and many different languages, as well as dialects within the languages. A full 75 percent of the army officers were of German stock, but only 25 percent of the army spoke German. The army of Austria-Hungary was thus composed mostly of soldiers often unable to understand orders and unwilling to serve their essentially foreign masters. Fearing a Slavic uprising within its borders, Austria-Hungary was anxious for a chance to teach a lesson to the country that seemed guilty of causing the unrest: Serbia. Underlying everything was the fact that many of the people of Europe were dissatisfied with their governments and with their aristocracies.

On June 28, 1914, the Archduke Francis Ferdinand of Austria-Hungary and his wife, Sophie, were riding in a motor parade through Sarajevo, the capital city of the empire's province of Bosnia-Herzogovina. A nineteen-year-old citizen of neighboring Serbia fired two bullets into the Archduke and his wife, killing them both. With that act, the young Serb felt that he was removing the obstacle that stood between the unity of all Slavs. Exactly one month later, Austria-Hungary declared war against Serbia. Russia began to mobilize on the German border, and the Germans requested it to stop. When Russia refused, Germany and Russia declared war on July 31.

Events moved rapidly after that. France had a treaty with Russia and, anxious for revenge, joined Serbia and Russia in war against Germany and Austria-Hungary. England with its secret military agreement with France and Russia, was also drawn into war, declaring their resolve on August 4, 1914. Italy declared itself neutral although it had a treaty with Germany and Austria-Hungary.

France, England, and Russia against Germany and Austria-Hungary was only the beginning. Later, the Allies included armies from England, France, Russia, Japan, Italy, Serbia, Portugal, Canada, Australia, and the United States. The Central Powers consisted of armies from Germany, Austria-Hungary, Bulgaria, and the Turkish Empire. The war raged for four years and four months. Historians say that the war ended on November 11, 1918, when the last gun was fired, but the Treaty of Versailles was not signed until June 28, 1919, five years to the day after the assassination of the archduke.

Turkey and Greece fought on until September 1922, when the Turks finally deafeated the Greeks. The French, Italians, and Russians were disgusted with their governments and were ripe for a new political movement. The war from 1914 to 1918 left twenty-six million dead and twenty million wounded.

Several new technological improvements in warfare resulted from World War I. Among them were the submarine, the airplane, and the steel helmet. The steel helmet was the only new, purely defensive war weapon developed in the war. It is impossible to determine the number of lives that were spared simply because of the steel helmet, but its significance was proven by its acceptance as standard equipment of every major army in the world. As helmet specialist Robert Oswald said, "Of the steel helmets worn by soldiers in World War One the German model, possibly because of its unique 'coal bucket' design, is the most easily recognizable. It has become famous to the extent that its origin is known in even remote areas of the world. Whatever the reason, the German steel helmet became the most significant symbol of World War One."

TWO

The Preliminary Model of the First Stahlhelm

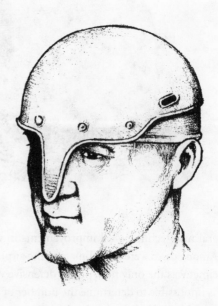

IN 1915, THE SO-CALLED Army Group Gaede (named after the commanding general), composed largely of Landwehr brigades (state reserve forces; for example Bavarian, Prussian, etc.), was stationed on the Vosges front, which was relatively quiet except for artillery exchanges. The nature of the terrain was such that since most of the front-line positions were faced with rock or built into rock, there were many wounds from rock splinters as well as from shell fragments. Thus, the need for usable and efficient head protection became imperative.

After prolonged unsuccessful efforts to obtain a helmet from the war ministry, the Army Group Gaede undertook extensive experiments to develop a helmet of its own. This was done at the artillery workshop of the army group in Muhlhausen, and development and production were overseen by Oberstleutnant Hesse. About fifteen hundred helmets were produced under his supervision.

The high-grade steel helmet was issued to troops stationed in front-line positions in naked rock areas where splinter wounds were to be expected. A skull-cap of cloth and leather was worn underneath it so that the heavy helmet might be worn more comfortably. The first concern in designing the Stahlhelm was to guard the front of the head, and the future "brow plate" would carry on that philosophy. With the introduction of the regular steel helmet in 1916, Stahlhelm M-16, the Gaede helmets were withdrawn and melted down.

THE FIRST HELMETS

THE FRENCH HELMET M-15

World War I was almost six months old when the French army started to fight with new headgear. The traditional soft felt kepi hat was replaced with a steel helmet for combat, and the vivid coloring of the French uniform was modified to grey-blue battle dress.

The new helmet was made of sheet metal .7 mm thick. A visor was welded on and extended around the complete bowl. A vent hole provided air passage at the top and was covered with a metal trim piece known as a "comb." This design, along with the comb trim, was unique and therefore easily recognized.

Five factories began manufacturing these helmets with the help of three thousand employees. Two million were produced in a span of six months. In

time, the entire French army was equipped with them. Since this helmet was first introduced in 1915, it is known as Model 1915, or M-15.

THE ENGLISH HELMET M-15

England was the second country to start using steel combat helmets. The English M-15 was first seen in late 1915, only two months after the French models appeared. It was more simple in design than its French counterpart. It appeared smaller, and the pan-shaped dome fluted out with a small visor that extended the entire circumference of the helmet. It looked the same from all angles. This exact design was later adopted by the United States. The M-15 bore a striking resemblance to the individually acquired helmets that the English army wore during the battles at Crecy and Agincourt during the Hundred Years War of the fourteenth and fifteenth centuries.

THE GERMAN HELMET M-16

Although Germany was the third country to begin using steel helmets, it may have been the first to begin design work on the drawing board. In some respects, it was an advantage for Germany to have a chance to test both the English and French types before making conclusive decisions of its own, judging that both designs were far from adequate.

After the final arrangements were made, the first German helmets were tested in mid-November of 1915 at the Kummersdorf Proving Ground. They were accepted well by the test personnel, and thirty thousand were ordered for immediate delivery. Once these helmets were issued to the troops at Verdun, the number of traumatic head injuries began to drop significantly. The formal introduction of the German helmet in 1916 marked the beginning of the Stahlhelm.

Helmet production commenced in the early spring at Eis Enhuttanwerk in Thale am Harz. Ten factories joined in the effort to manufacture helmets, and together they produced from 3,500 to 4,000 per day. The total output at the end of the war was 8.5 million helmets.

The man responsible for the design of the German Stahlhelm was Dr. Friedrich Schwerd of the Technical Institute of Hanover. In early 1915, Schwerd completed a study of head wounds that were the result of trench warfare. He was working with the medical department of the Eighteenth Army Corps and was captain in the artillery. He was a staff member of the communications Zone Inspectorate 11. He forwarded a recommendation for steel helmets through Professor Dr. August Baird, a consulting surgeon to the Eighteenth Army. He was ordered to Berlin shortly thereafter. Schwerd then undertook the job of designing and producing a suitable steel helmet.

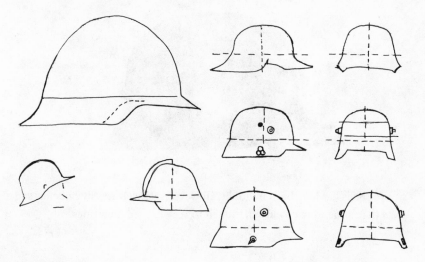

Copies of Schwerd's original pen sketches. The original examples are in the Armed Forces Archives, Potsdam. Note that the last two sketches include the side lugs for the frontal shield. This indicates that the armored frontal shield was intended from the very beginning and not added later, as some thought. Also, note that the chin strap lugs are on the outside of the helmet. These must be the last of the proposed designs, because the issued helmet differed from these sketches only in the subsequent internal placement of the chin strap lugs.

THE M-16: DESCRIPTION OF THE FIRST STAHLHELM

The design of the completed helmet included three sections: the dome, the visor, and the neck guard. The dome was the main head covering and was cylindrical in shape and somewhat flat on top. The 6½ inch visor extended out 1⅜ inches to provide shade and protect the wearer against bad weather. It also acted as an open shield against fragments. The 2½ inch neck guard flared out below the brim, which provided more protection around the neck and ear area.

The complete m-16 weighed two pounds and six ounces. The color was field green, and the metal was composed of manganese, nickel, silicon, and carbon steel, which was often referred to as nickel steel. Its thickness was between .40 and .45 inches, and it is believed to have been pressed hot on electrically heated dies and later dipped into a mixture of japan for the antirust finish.

Helmets were referred to as "shells" when they were empty of any liners and straps. The m-16 shell was manufactured in six sizes: 60, 62, 64, 66, 68, and 70. This was the outside measurement directly below the rivet holes. For example, the bowl size of 60 would accommodate a head size of 50 and 51 centimeters. (This is equivalent to the American sizes of 6¼ to 6⅜ inches.) The existence of

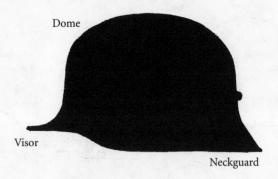

Dome

Visor

Neckguard

size 70 shells has long been in doubt. However, relatively recent evidence suggests that some were produced, though in very limited quantities.

Size	Fit (cm)
No. 60	50.5–52
No. 62	52.5–54
No. 64	54.5–56
No. 66	56.5–58
No. 68	58.5–60
No. 70	60.5–62

The sizes were engraved on the inside of all of the shells, and the manufacturer's identification could be found along with it. The inscription was on the left side of the skirt. The manufacturer's code letters appeared first. There were eight factories involved in the manufacturing, therefore, eight different code letters were represented in all. The head size was never marked in the shell, only the bowl size. It was the metal liner that offered the correct head sizes.

THE CHIN STRAP

All M-16 helmets were equipped with one style of chin strap. This applied only if the helmet were made in Germany for the German army. The strap was the same type found on leather spike helmets. It consisted of a strip of leather looped around two slide buckles and connected to each end by means of attaching "eyes." These were fastened to the inner side of the neck guard to specially mounted lugbolts. The eyes turned freely.

While the strap's duty was to hold the helmet on firmly, it was designed not to hinder the soldier who wore it. The helmet was meant to fit securely without the chin strap pressing too hard against the neck. The proper length adjustment was applied by simply sliding a buckle. Since the straps were detachable, many

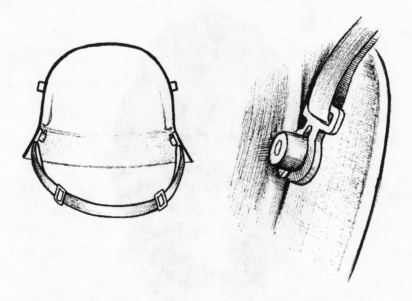

were lost, and so replacements were commonly provided in warfare. The Austrians produced replacement straps of woven or sewn cloth with rectangular attaching eyes that prevented them from falling off the lugs of the steel helmet.

THE LINING

The lining inside the m-16 helmet offered a cushion and the glove-tight fit necessary for wearing steel headgear. This consisted of a mounting band of leather or steel that ran the length of the inner wall of the shell. It was attached to the shell by three rivets, one in the back and one on each front side. Leather bands were used for the early helmets and steel was used for the later pieces when a leather shortage forced improvisation.

Three pads were attached to the mounting band and were folded upward within the dome of the helmet, and each was backed by a cushion. One of the pads was located at the forehead and the other two were positioned against each side of the head. Each tab contained a pocket that included a small mattress filled with horsehair or other material. This mattress was kept in position by means of cloth tapes that could be tied.

A leather string ran through each of the three pads by means of a small hole. By adjusting it in the same manner as a drawstring, the paddings could be adjusted to fit the head like a cap. Actually, the pads were cushioned evenly on the crown of the head and allowed a clearance between the inner top of the crown, so as to distribute a hit or blow as evenly as possible over the head surface. Thus, the scalp and the top of the head could still receive a supply of blood freely. With

13

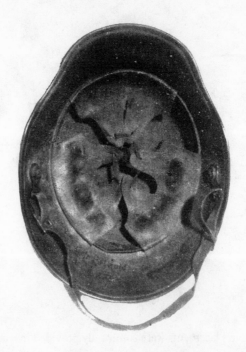

the three spaces between each cushion, the M-16 liner included an open passageway, so, in the case of denting the helmet on the sides or top, the protected soldier avoided contact because of that space between the helmet and head.

In addition to the comfort and safety designed into the M-16 liner was the ease of changing the fitting. If the wearer felt that the supporting cushions were too hard or thick, he was at liberty to remove some of the stuffing to whatever degree he pleased. Or, if the helmet fit his head too loosely, he merely had to loosen the strings of the enclosed pads and pack the needed amount of any stuffing available. The stuffing could be any type of material, such as burlap, cloth, a layer of cotton, or wool. Many times scraps of paper or bits of grass were used.

The complete liner weighed 4½ ounces. The leather used was dark brown in the earlier models, however lighter and better quality materials were found in later issues. There was no size selection with the M-16 liner since the adjustable fitting was provided in the three-pad feature; therefore, no size markings can be found on the leather liner.

SIDE LUGS

No other feature is as recognizable on the M-16 as the side lugs. They stand out and are quite impressive. The spindle-like bolts that stick out from each side

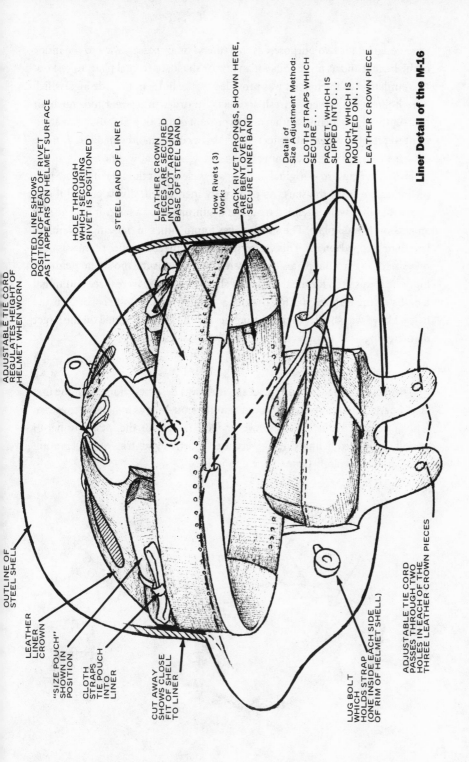

ADJUSTABLE TIE CORD
REGULATES HEIGHT OF
HELMET WHEN WORN

DOTTED LINE SHOWS
POSITION OF HEAD OF RIVET
AS IT APPEARS ON HELMET SURFACE

HOLE THROUGH
WHICH SECURING
RIVET IS POSITIONED

STEEL BAND OF LINER

LEATHER CROWN
PIECES ARE SECURED
INTO SLOT AROUND
BASE OF STEEL BAND

How Rivets (3)
Work:

BACK RIVET PRONGS, SHOWN HERE,
ARE BENT OVER TO
SECURE LINER BAND

Detail of
Size Adjustment Method:

CLOTH STRAPS WHICH
SECURE . . .

PACKET, WHICH IS
SLIPPED INTO . . .

POUCH, WHICH IS
MOUNTED ON . . .

LEATHER CROWN PIECE

Liner Detail of the M-16

OUTLINE OF
STEEL SHELL

LEATHER
LINER
CROWN

"SIZE POUCH"
SHOWN IN
POSITION

CLOTH
STRAPS
TIE POUCH
INTO
LINER

CUT AWAY
SHOWS CLOSE
FIT OF SHELL
TO LINER

LUG BOLT
WHICH
HOLDS STRAP
(ONE INSIDE EACH SIDE
OF RIM OF HELMET SHELL)

ADJUSTABLE TIE CORD
PASSES THROUGH TWO
HOLES IN EACH OF THE
THREE LEATHER CROWN PIECES

15

were designed for two purposes: as "breathers" or air passageways to the inside of the helmet and as pivot bolts for an armor shield, or frontal plate, to rest on.

Ventilation of the helmet was provided by the holes in the side lugs, which were hollow and served as fresh air ducts. In order to prevent too much air from entering in stormy weather and to prevent disturbing sounds, it was sufficient to plug the holes with paper, mud, or any other handy material. If more air was needed, a simple cleaning of the lugs was all that was necessary.

From Dr. Schwerd's original drawings it is clear that the side lugs were part of the design from the start. As previously mentioned, this is powerful evidence that they were intended from the beginning to be a fastening device for an add-on frontal plate. The side lug was found either in one single diameter length or stepped length. This varied with the helmet size, since there was only one size of the frontal plate. There were three lengths for the single diameter lug, which was used for the three larger shell sizes. There were three additional lengths in the stepped lug for the three smaller sizes. In other words, if the helmet bowl was smaller than the thickness of the shield, it rested on the larger diameter lug as in the case of the four smallest sizes.

THE FRONTAL PLATE

The frontal plate was a forehead shield. It was designed along with the first M-16 helmet and they were produced simultaneously so as to be used together. Only 3 percent of the shields, however, were issued for the total number of initially produced helmets. They were used mostly by sentries on observation posts, and particularly by machine gun units.

Due to its shape, the frontal plate fit all helmet sizes. It was hung on the side lugs and secured on the steel helmet by means of a leather strap. The strap was then seated against the rear rivet that held the lining. Because of its heavy weight, the plate could only be used for short periods of time. The plate weighed five to seven pounds and was .23 to .25 inches thick. There was only one size available.

The frontal plate protected against enemy fire at distances beyond fifty meters. It extended over the forehead and temples and overlapped the helmet far enough to protect the back of the head from bullets striking from the front. The frontal plate, along with the rest of the helmet, could lose its strength and ballistic value if damaged by fire. Those who wore the plate almost always used additional armor, such as a breast plate or other types of body armor.

There was a different type of rivet head in the back of the M-16, compared to the two side rivet heads. This was to provide a fitting for the frontal plate's strap. The two side rivets were rounded flat and fit close to the body. The rear rivet protruded ¼ of an inch from the helmet and allowed a more snug fit by keeping the leather strap of the shield from slipping upward.

The specially applied lacquer coating found on the surface of the M-16 helmet protected it against rust. If the helmet were damaged, the owner was instructed to either touch up the spot or have it repainted as soon as possible. When the German soldier was not in need of his helmet in transport or marches and not in combat or other dangerous situations, the helmet was not worn and was buckled on the knapsack cover or pocket in the rucksack. Later, a special device was issued to hang the helmet on the belt behind the bayonet.

The shell's finish became shiny with much use, which was an unfortunate and dangerous condition. So in order to eliminate attention from the enemy, a cloth cover was made and issued. This was important to those soldiers on patrol and observation missions. There were about 800,000 covers made and distributed, which figured to about one cover for every ten helmets. The cover was fastened to the helmet by means of a drawstring that was tied on the back at the center of the neck guard. Cutouts were provided in the cover for the side lugs.

Although the Austrians also manufactured covers, they did not use them as widely as the Germans. Many had leather or oilcloth washers around the vent lugs.

There were variations in color for the M-16, but all were about the same greenish grey. The primary difference was that the early helmets were finished with more lacquer which gave them a harder finish and a shinier appearance. The later ones had a softer-looking finish and were a lighter grey mixed with the green.

While camouflage types seemed to follow no single pattern, they did fall into four different types: patch, splotch, zig-zag, and tortoise shell. The patch, which seemed to have been more frequently used, had large areas of green, maroon, brown, and yellow outlined in broad black lines. The splotch type had no outlining and was generally far more effective for camouflaging. The zig-zag was used for the forest and for experimental purposes. Another more rare form was the tortoise shell. The tortoise shell had narrow black lines enclosing dark greens and browns on the front and sides and yellow in the back, possibly for recognition.

Insignias were not authorized to be worn on the helmet during WWI except for several small special units, including imperial guards. This general prohibition applied to badges, decals, or painted designs of any kind. The only added materials to M-16 helmets were paint for camouflaging, a cloth cover, or the frontal plate. Therefore, about 75 percent of the helmets were free of any add-ons.

Patch

Zig-zag

Tortoise shell

Splotch

Normally, the steel helmets were packed in sized crates. The Austrians, for example, typically shipped their helmets in single crates of fifty each. Stenciled on each crate was the number of steel helmets enclosed along with their size and manufacturer. Also, there were three information pamphlets placed on the inside of the cover. Fifty crates with 2,500 helmets could be loaded in one standard boxcar.

Because of the urgent need for the M-16 on all the fronts, they were initially shipped by separate size groups in entire boxcar loads from the German interior, with the distribution of the size classes handled at the front. According to the normal head sizes, the greatest concentration of helmets supplied was in sizes 64 and 66. Of these sizes, 40 percent of each were produced; only 10 percent of the smaller and larger sizes 62 and 68 were produced.

Steel helmets that did not completely meet the prescribed ballistic conditions of a 13 gram pellet with 240 meter per second muzzle velocity were judged unsuitable. They were classified as inferior for field use and marked with a red stripe on the inner side of the neck guard. Such helmets were not issued for combat, but they were utilized for training programs in replacement units of technical troops and other noncombat formations. The containers used for packing the unsuitable M-16 helmets were also clearly marked with a red horizontal stripe for proper recognition.

Each German soldier was issued one M-16, which then belonged to him. As for the other equipment, there was a repair and supply depot set up in a motorized fashion in the combat area for small repairs of the helmet. All other helmets that became unusable were shipped immediately through war material collecting points of steel helmet repair shops. Those from all fronts north of the Danube River went to Brunn at the Gottlieb Brothers' helmet factory, and all fronts south of the Danube were sent to Cilli at the A. Weston helmet factory.

A VARIANT STAHLHELM: THE HIGH VISOR HELMET

There were two visor clearance types for the M-16 helmet. The size of the visors varied by a quarter of an inch. The difference did not seem impressive, but it was very apparent and gave a different look to the helmet. It is believed that the lower visor was a later helmet, although vision was limited, which is ironic, because the logical sequence would have been the other way around. It is believed by some that the higher visor models were the exact first models that circulated and that they were almost a prototype, or an experimental helmet. Whether true or not, the high visor is not as plentiful as the regular low visor helmet is now. In fact, it is a rare find.

A high visor may be determined from the ordinary by placing the two in front of each other with one visor pointed directly at the other. The high visor will extend over the regular helmet's visor if they are both on the same level.

A VARIANT STAHLHELM: THE FULL VISOR HELMET

Another helmet variant appeared during the war when the full, or high, visor German helmet was produced and issued to troops. It was the same in all respects to the M-16, except the front visor extended from the dome at a much higher location. The flare-out in this area was almost as great as that of the neck guard. The length of the visor from the face remained the same; it was only the extra space by the forehead that was increased.

At the time, officials believed the full visor, sometimes called the "new" helmet, would eventually replace the regular M-16. It may have been designed for easier gas mask use or for greater ballistic deflection. In any case, it vanished as fast as it had appeared. Since this variant was manufactured for such a short period, it has become extremely rare. Data concerning production and distribution remain elusive.

STAHLHELM M-17/18

Subsequently, another Stahlhelm was introduced, the M-17. Representing an improved model, it is sometimes labeled the M-17/18. There had been complaints that the M-16 strap lugs were falling off from damage occurring in

The Full Visor helmet.

battle. Sometimes the strap could not be loosened, or disengaged altogether, in a quick manner. Many times the strap would fall off and be lost. The M-17/18 provided a solution: a different type of strap with different connections and the elimination of the strap lugs. The leather strap was attached to a "D" ring connection that was fastened directly to the liner band. Also, the grade of leather found on the M-17/18 was of better quality than the former M-16.

The M-17/18 had an adjusting buckle and hook-up type buckle that helped keep the chin strap at the correct position and prevented the slipping of the adjusting buckles, which occurred with the M-16. Since there were no strap lugs on the M-17/18, the rivet showing directly out and opposite this site on the M-16 ear guard was absent. The color of the shell remained the same. The side

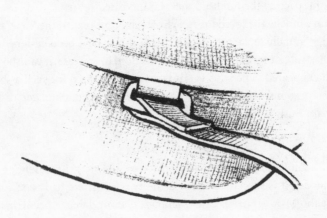

vent, liner detail, size, and manufacturers' inscriptions did not change. The only improvement made was to the strap and strap connections.

Most of the documentation on the production of WWI German helmets after the 1916 model were destroyed in 1945. Details of development and production are, to this point, unavailable.

THE SPECIAL CUT-OUT HELMET: M-18

Two years after the M-16 was issued to German troops, another form of the helmet appeared. It was known as the "special" or "cutout" helmet and is now often labeled the "cavalry" model. It was exactly like the M-16 in most respects. The one exception is that there were no inside connecting lugs for the chin strap and the different strap fit directly to the liner band. Additionally, there was a cutout area on both sides of the neck guard where the ear was located.

The purpose of the cutout helmet was to provide better hearing in the trenches, to facilitate use of field telephones, and to reduce explosive resonance. The German soldier complained that many times his hearing was hindered and danger prevailed because of the neck guard shielding his ears, especially in trenches. German technicians experimented on a model by trimming out a section, allowing a curve in the neck guard. It was enthusiastically accepted by the German staff, and it is said that thousands were manufactured. Despite the general assumption that the model went into mass production and that many survived the war, this helmet is relatively rare today.

THREE

The Stahlhelm in Other Countries

THE TURKISH STEEL HELMET

Germany made virtually all of its own military supplies in WWI, and its allies relied on it to provide a great deal of equipment. One of the badly needed items was steel helmets, of which Germany delivered 5,400 to Turkey. They were manufactured in Germany and designed from the M-16, but the front visor was missing, and this was understood by the Germans to be for religious reasons.

There was a total of 2,850,000 soldiers mobilized in Turkish forces, but only one out of five hundred wore a steel helmet. Perhaps it would have been better to outfit more, since 325,000 were killed and 400,000 were wounded.

The Turks are known to have rejected any more than the 5,400 delivered, and the reason for their reluctance to accept more is not clear even today. Therefore, an unknown number from the overrun ended up being issued to the somewhat puzzled German armed forces. Those mysterious helmets worn by German forces late in the war have consistently been misidentified in photographs from the period.

THE AUSTRIAN STAHLHELM

Since Austria-Hungary was one of Germany's allies in WWI and a partner in the military effort, their steel helmets are included along with the Stahlhelm's evolution. They began using M-16 helmets in WWI when the war was in its second year.

The Austrian M-16 was almost identical to the German in appearance. It can be identified quickly by the location of the chin strap rivet head, found higher up, closer to the side lug. The Austrians manufactured their own helmets in addition to others they purchased from Germany. It had the same dome shape, visor, and ear guard skirt as the German model. The differences included paint

SIZE	FIT IN CENTIMETERS	PRODUCTION RATE (%)
No. 62	52.5–54	10
No. 64	54.5–56	40
No. 66	56.5–58	40
No. 68	58.5–60	10

Germany manufactured 486,000 helmets that were issued to Austrian troops. Both countries repaired and repainted each others' helmets picked up from the battlefield. Thus, it was possible to find many different combinations of parts in any WWI Stahlhelm.

THE BROW PLATE AND STRAP FOR THE AUSTRIAN M-16

Since the Austrian M-16 helmet was designed to accommodate a brow plate, side lugs were planned from the beginning to be used for this reason, similar to the German M-16. The frontal plate was produced in only one size for all four numbers of helmets. The production of the shields was 3 percent of that of the helmets. Each soldier had a helmet of his own, but the frontal plates and covers were kept in company equipment stores and issued as was deemed essential.

The frontal plate was suspended on the side lugs of the helmet. It was held tightly about the helmet with a cloth strap sewn to the shield, not a leather

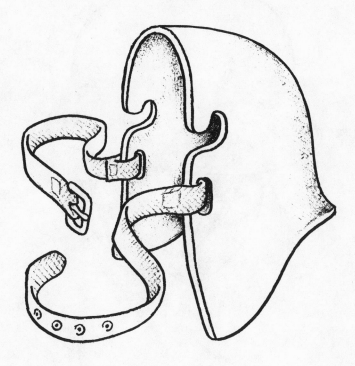

strap like the German models. The shield protected the head against infantry bullets fired at a distance of at least fifty meters. This shield was heavy and worn on special occasions only, such as in observation posts and at special guard positions. The thickness of the shield was five millimeters. All Austrian frontal shields were painted field brown.

THE AUSTRIAN BERNDORFER HELMET

Besides the regular Austrian M-16 trench helmet, another Stahlhelm of Austrian origin with a very different design was used. This was the Berndorfer M-17 which appeared in early 1917. Made by Berndorfer Metallwarenfabrik Arthur Krupp A.G. at Berndorf, in lower Austria, this helmet was used only by the Austrian army and was made in very limited numbers.

The characteristic features of the Berndorfer were unique, and it claimed a style all its own. It had no resemblance to any of Germany's helmets or to any helmets from any other nation in the world. The dome was shorter in height and appeared rounder on top. The front visor had a shorter extension and flared out in a square manner. A "crimp" was distinguished at the beginning of

the ear guard area located on each side of the visor. The ear guard showed a very small dip from the visor portion giving the Berndorfer more of an overall bowl-like appearance than the German model.

The Berndorfer was issued with three different liners similar to the German ones, but they had metal eyelets in the tab holes. The chin strap was made of a single piece of cloth with slide buckles. There were no side lugs on the Berndorfer M-17. Instead, there was one lug on the very top. It was a round, flat piece that sat on an extension that housed four holes. It had the same purpose as helmets with side lugs: for ventilation and as a holder for a frontal piece. The size of the bowl was inscribed almost to the rear of the inner wall of the skirt. All Berndorfers were painted field brown; and since it was used by both the army and navy, there were no insignias applied.

This helmet took a tan cloth cover to eliminate light reflection. When not used, it was packed in a cloth bag. The cover included a leather section with holes at the top to allow for the vent lug to extend. Some 140,000 were issued and used until the end of the war. They survive in small numbers and are considered rare.

Brow Plate for the Berndorfer. The brow plate for the Austrian Berndorfer was a completely different conception than the shield that fit the regular issue

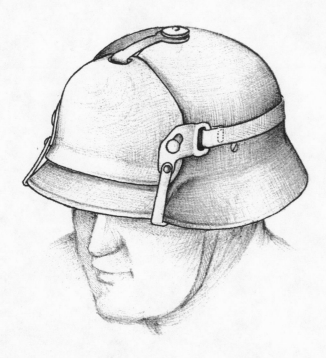

Austrian м-16 helmet. This was due to the irregular shape of the Berndorfer helmet itself and the form that the plate had to fit. Also, the necessary fittings were altogether different.

The plate appeared smaller because of the dome's roundness. It was held to the helmet in three sections. First, it was secured from the plate where two straps extended down and attached to each side of the visor at the crimp area. Also, there was a strap that wrapped around the crown and secured to each side of the plate. Finally, the top of the plate was held rigid with the support of the round, top vent fixture.

All straps were cloth and colored brown, with the brow plate painted field brown. Only about 3 percent of Berndorfer helmets were equipped with brow plates, so they are extremely rare.

FOUR

The Peace Treaties and New World Map of 1919

THERE WERE SEVERAL peace treaties drawn up at the conclusion of WWI, and the suburbs of Paris were the scene for five of them. The most famous was the German document, which was negotiated at the former palace of the French kings and became known as the Treaty of Versailles. The Hungarian treaty was signed at the Grand Trianon, in the park of Versailles and the Austrian treaty was signed at St. Germain, the Bulgarian treaty at Neuilly, and the Turkish treaty at Sevres.

The five Central Powers—Germany, Austria, Hungary, Bulgaria, and Turkey—were individually dealt with at the peace table. The treaties were countersigned by thirty-two allied nations' representatives (thirty-one in Germany's case, China abstaining). The German agreement dealt with four main points: loss of German territory, destruction of German military power, distribution of German colonies, and Germany's acceptance of the blame for starting the war and thus agreeing to make reparations. (Reparations were so punitive that they amounted to the economic destruction of Germany. They were subsequently drastically scaled down, largely due to the analyses of British economists and the United States government.)

THE LOSS OF TERRITORY

Germany. Germany lost Alsaçe-Lorraine to France, Eupen and Malmedy to Belgium, and parts of Schleswig to Denmark. Germany also lost Danzig to free city status; Sudetenland and a strip that became known as the Polish Corridor, which gave Poland access to the Baltic Sea while splitting Germany from East Prussia, and a considerable piece of real estate to a reemerging Poland. Poland had disappeared from the map in 1795, having been divided up among Germany, Austria, and Russia.

The German steel helmet was respected by all countries, and many features were adopted in other helmet designs in later years.

Austria. Austria was transformed to a land-locked nation of less than seven million people.

Hungary. Hungary separated from Austria and was also reduced to a land-locked nation with the loss of two-thirds of its former population.

Bulgaria. Bulgaria lost its influence over the Dardanelles when it lost the territory it had gained in the Balkan Wars in 1912 and 1913.

Turkey. Turkey lost its territories to the newly created entities of Syria, Iraq, Lebanon, and Transjordan which were under the control of France and Britain.

Russia. Russia lost a total of land greater than all the Central Powers except Turkey. The loss possibly served as a punishment for signing an early treaty with Germany. Still, Russia lost less territory under the later Allied peace treaties than it did under the treaty Germany drew up upon Russia's original surrender in December 1917.

Nations reappearing on the European map or newly created as a result of the 1919 treaties were: Poland, Finland, Lithuania, Latvia, Estonia, Ukraine, Czechoslovakia, Armenia, Georgia, Azerbaijan, and Yugoslavia.

DESTRUCTION OF GERMAN MILITARY POWER

The following were turned over to the Allies: 1,700 airplanes; 5,000 locomotives; 25,000 machine guns; 2,500 field guns; 2,500 heavy guns; 6 battle cruisers; 10 battleships; 50 destroyers; all the submarines; and much of the other

Casualties of All Belligerents in World War I

Country	Total Mobilized Forces	Killed and Died	Wounded Casualties	Prisoners and Missing	Total Casualties	Per cent
	Number	Number	Number	Number	Number	
ALLIES						
Russia	12,000,000	1,700,000	4,950,000	2,500,000	9,150,000	76.3
France	8,410,000	1,357,800	4,266,000	537,000	6,160,800	73.3
British Commonwealth	8,904,467	908,371	2,090,212	191,652	3,190,235	35.8
Italy	5,615,000	650,000	947,000	600,000	2,197,000	39.1
United States	4,355,000	126,000	234,300	4,500	364,800	8.0
Japan	800,000	300	907	3	1,210	.2
Romania	750,000	335,706	120,000	80,000	535,706	71.4
Serbia	707,343	45,000	133,148	152,958	331,106	46.8
Belgium	267,000	13,716	44,686	34,659	93,061	34.9
Greece	230,000	5,000	21,000	1,000	27,000	11.7
Portugal	100,000	7,222	13,751	12,318	33,291	33.3
Montenegro	50,000	3,000	10,000	7,000	20,000	40.0
Total	42,188,810	5,152,115	12,831,004	4,121,099	22,104,209	52.3
CENTRAL POWERS						
Germany	11,000,000	1,773,700	4,216,058	1,152,800	7,142,558	64.9
Austria-Hungary	7,800,000	1,200,000	3,620,000	2,200,000	7,020,000	90.0
Turkey	2,850,000	325,000	400,000	250,000	975,000	34.2
Bulgaria	1,200,000	87,500	152,390	27,029	266,919	22.2
Total	22,850,000	3,386,200	8,388,448	3,629,829	15,404,477	67.4
Grand total	65,038,810	8,538,315	21,219,452	7,750,919	37,508,686	57.9

military equipment. In addition, 64,000 steel helmets were destroyed. For the future, Germany was to be allowed 6 battleships, 6 light cruisers, and certain small vessels, but no submarines. The army was to be limited to 100,000 men, including officers, and the manufacture of munitions was to be restricted.

DISTRIBUTION OF GERMAN COLONIES

In Africa, Togoland and Cameroon were to be jointly administered by Britain and France. German East Africa (Tanganyika) was assigned to Britain, and German Southwest Africa to the Union of South Africa. South Pacific possessions were to be administered by Australia and New Zealand (part of New Guinea, the Bismark Archipelago, Western Samoa). North Pacific possessions were handed over to Japan, including Guam, Saipan, and Tinian in the Marianas Islands; Kwajalein and Eniwetok in the Marshall Islands; and Truk and Palau in the Caroline Islands. These would later become household names in the United States during the Pacific war with Japan.

GERMANY HAD TO ACCEPT THE BLAME FOR THE WAR

Article 231 of the Treaty of Versailles stated that "the Allied and Associated Governments affirm and Germany accepts the responsibility of Germany and her allies for causing all the loss and damage to which the allied and associated governments and their nationals have been subjected as a consequence of the war imposed upon them by the aggression of Germany and her allies." Article 231, together with unrealistic reparations demands, became a major source of aggravation that Adolph Hitler later exploited so effectively and with such devastating results.

FIVE

Postwar Unrest

WITH THE END OF WWI came questions such as "What happened to the Central Powers?" and "What types of governments followed?"

TURKEY

The prewar goals of the Turks again manifested themselves, led by a talented modernizing army officer, Mustafa Kemal (Ataturk). In 1922 the old Ottoman Empire was dissolved and the new Republic of Turkey was established along the parliamentary lines of the West. With Kemal at the head of the new government, Turkey's capital was moved from Istanbul inland to Ankara, and there followed a new peace settlement with the Allies in the Treaty of Lausanne. This involved a new principle: the transfer of populations. The Lausanne treaty of 1923 was more advantageous to Turkey than the abrogated treaty of Sevres of 1919. Turkey gained the respect of the world and remained neutral in WWII. In fact, the brief, twenty-year period from 1919 to 1939 saw more vigorous social reform than any other comparable 100 year period in Turkish history.

AUSTRIA

Until 1938, Austria was first under the leadership of Engelbert Dollfuss, who was followed shortly thereafter by Kurt Schussnig. Dollfuss was assassinated by the Nazis in 1934, and Schussnig was imprisoned by Hitler in 1938. The crime of both Austrian leaders was democratic service, avoiding the two extremes of communism and fascism. Austria was subsequently absorbed into Hitler's Reich.

HUNGARY

The new nation got the leadership the Allies wanted by the peace treaty of Trianon in June 1920, confirming the loss of Transylvania. The actual event caused the liberal Count Michael Károlyi to resign, and a Communist Party

Helmets with various painted insignias such as a skull and crossbones or potato masher grenades were Frei Korps insignias for the various organizations that sprang up after World War I.

chief, Bela Kun, staged a coup. The allies allowed Romania to invade Hungary in 1919 and depose the Communists. The Romanians withdrew in 1920, and a French-inspired government took over that was monarchist in character and later pro-Nazi. Therefore, the loss of Transylvania in 1920 helped move Hungary out of the democratic and into the fascist camp.

BULGARIA

As with Germany, Austria, Hungary, and Turkey, the period of so-called "peace" following the treaties of 1919 was largely one of internal revolution, conspiracy, intrigue, and murder as Communists, fascists, monarchists, and moderates fought to control the government. From 1920 to 1923, a fairly popular government under Alexander Stambolisky ruled, but was overthrown by authoritarian King Boris, who was himself thrown out of power by fascists in 1934. Their reign was put to an end by a counterrevolution by King Boris, who imposed a royal dictatorship that lasted into WWII.

GERMANY

The Germans adopted a new constitution and elected Friedrich Ebert as president in 1919. Germany became a republic, but when the new government signed the Versailles peace treaty, democracy virtually died at birth. The Left and the Right were both outraged that a German political group signed a treaty calling for the dismantling of Germany. In 1920, a renegade military "putsch" attacked the democratic government for a few days, but Ebert put it down with a general strike and a show of counterforce. The Communists went on a rampage in the

The captor is wearing a special cutout.

A Frei Korps wearing painted insignias on the fronts of their helmets.

A gun position against revolution. Note the Turkish visorless helmet on the far right.

Insignia example of a Frei Korps, the upright swastika.

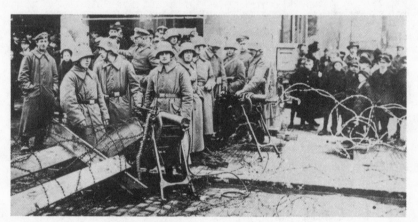

Variations of helmets left over from WWI.

A white helmet stands out among dark helmets.

demilitarized zone of the Ruhr, and Ebert sent troops in to put down the riots. France, considering the presence of German troops in the Ruhr as a breach of the Versailles treaty, invaded Germany and occupied the Ruhr and Frankfurt during April and May 1920.

The elections of June 1920 sounded the death knell for the new republic. The Versailles treaty, the government which signed it, and the French invasion while the democrat Ebert was chasing Communists brought the right-wing vote out in strength. The Communists, feeling they had French and Allied support, also turned out at the polls. Both extremists won victories, and the moderates lost. Ebert was temporarily safe, elected in 1919 for a period of seven years. In April 1921, the Allies presented their reparations bill to Germany, over the protest of prominent and responsible Allied economists. Walter Rathenau, the German minister for reconstruction, who was a Jewish democrat, took the moderate course that Germany should at least try to meet the possible demands, while both the Left and the Right favored outright rejection. Rathenau was supported by another moderate, Matthias Erzberger, leader of the Catholic Center Party. He was one of the men who signed the Armistice and who was the number-one target of the right wing. He was assassinated in August 1921.

Erzberger was the first victim of the Versailles treaty and Rathenau was the next, succumbing to bullets in June 1922. It can be assumed that the men were victims of right-wing outrage, but the left wing was equally capable of assassination. A further boost for the extremists in the reparations bill of 1921 was the loss of more German territory. Germany lost Upper Silesia, which extremists felt Rathenau condoned.

The so-called "Polish Corridor" that split East Prussia from the rest of Germany gave Poland access to the Baltic Sea. It was to become one of the major conflicts leading to WWII. As a footnote, the Polish Corridor issue was nothing new to either the Poles or the Germans. First occurring in 1386, it remained in existence for many years. Territorially speaking, as a result of the Treaty of Versailles in 1919, the Germans lost the regions of Malmedy and Eupen to Belgium, and part of the district of Schleswig to Denmark, and Prussia was split once again.

Immediately following the peace of 1919, the warring nations found a huge civilian force of ex-soldiers trying to fit back into a normal society. Not even a majority of the six million men mobilized for military duty during WWI had killed anybody, but all of them were psychologically prepared to kill. Millions of veterans found it difficult to settle down to regular jobs, and before long even the boring jobs were difficult to find. The emotions that arose out of desperate necessity were not easy to curb, especially among the vanquished. The veterans, who were bored with civilian life, felt betrayed by their government's casual concern about unemployment and adjustment problems.

Thus, the situation was ideal for the formation of radical veterans' organizations in Germany, as well as throughout the world.

In politics, the mood was ripe for aggressive talk and action, and politicians were quick to read the signs and act accordingly in order to thwart revolution. Young Communists, inspired by the momentous Communist coup in Russia, also found an ideal situation in the immediate years of postwar European unrest, boredom, discontent, and unemployment. Monarchies crumbled and governments reformed to a more democratic pattern. "Privileges or discrimination due to birth or rank as recognized by law are abolished," read the new German constitution that was drawn up in 1919. Also known as the Weimar Constitution, the German document provided for a Reichsrat, the upper house of representatives, and a Reichstag, the lower house. The president was allowed liberal power, but he had to appoint his Chancellor from the Reichstag majority party.

The aristocrats were out of power in Germany. The military was insignificant. Unemployment and inflation were rampant. The mark, worth twenty-four cents in gold in 1914 was down in value to seven cents in 1919. In October 1923, authorities in Berlin ordered taxi cab drivers to multiply the amount shown on their meters by 100,000,000. A registered letter mailed from Berlin to New York City in 1923 required eight stamps with the prewar value $2.5 billion.

In 1920, in Bavaria, a right-wing ultra-patriotic party was formed, called the National Socialist German Workingman's Party (NSDAP), known in history as the Nazi Party from the first four letters of the German word for "national." With the German economy in utter chaos and extremists fighting for control of the government, Adolf Hitler was named chancellor in 1933. The appointment was completely legal under the Weimer Constitution, which, in fact, required it, because Hitler was the leader of the largest party in the parliament. However, the appointment was delayed by President Hindenburg simply because, as an aristocrat, he was opposed to turning the German chancellorship over to an Austrian rabble-rouser.

The Third Reich was established as Hitler destroyed the republic. The First Reich was the Holy Roman Empire of the fifteenth century. The Second Reich was created at Versailles in 1871 by Bismarck at the end of the Franco-Prussian War and was dissolved at Versailles in 1919. The Third Reich, as promised by the new German fuhrer, was to last a thousand years.

From 1933 to the start of WWII in 1939, Hitler built up the military, pushed his brand of social reform, and eliminated the Jews and Communists from German society. Besides his war with Communists and Jews, he purged Germany of all political opposition. This included his own corps of the Sturmabteilung (SA), former veterans who led Hitler's street fighting gangs during his rise to power. The Schutzstaffel (SS) became his tool for eliminating anyone who had a different agenda than his.

For the first time in modern German military history, Hitler's purge of aristocratic officers in 1937 took the military out of the hands of the Prussian Junkers made up of the landed gentry. On September 1, 1939, Germany invaded Poland, starting the Second World War. Following the annexation of Austria and the forcible incorporation of the Czech lands, the invasion was too much for the Anglo-French governments. But from September 3, 1939, when Britain and France declared war on Germany, neither side made a major military move on the ground until May 10, 1940. This was a period of roughly nine months. The world called it the "phony war."

SIX

Transition of the Stahlhelm

REICHSWEHR INSIGNIAS

Between the period from the end of WWI to the beginning of the Third Reich (1919–33), Germany was federal in structure and consisted of states (lander). During this time, the territorially organized German army displayed insignias on their helmets to represent the states. These insignias were painted on the left side of the helmet. The emblems were shield-shaped, with the bottom of the shield more pointed than the later Nazi national colors types. Only one insignia was used per helmet, with the exception of the helmet of special guards' units, which had one insignia on each side.

The emblems were discontinued when Hitler abolished the traditional states when he came to power.

THE TRANSITIONAL HELMET

After 1918 the Reichswehr retained the M-16 and M-17/18 in service. The chin strap rivets were removed from the M-16 models and the rivets were peened over. These helmets were marked with sizes followed by "aA," which indicated an old model. Markings on these converted helmets were, for example, "64 aA/54."

In 1931 a new liner was introduced that immediately replaced the old M-16 and M-17/18 liners and was used until the end of the war. There were a few small changes made to the liner throughout the years, but mostly for experimental purposes.

When Hitler began to mobilize his troops, they used the transitional helmet from the WWI blanks. However, there was a completely new 1931 liner and strap assembly and a new exterior paint color, as well as insignias placed on each to represent the service or organization the owner represented.

Other style models were introduced later, at which time the transitional helmet's use began to fade. Although there was no manufacturing of this helmet during WWII, many were used, having been left over from the early 1930s.

Reichswehr Insignias

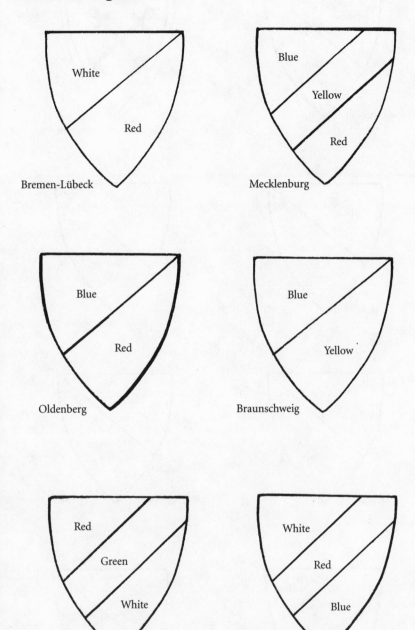

Bremen-Lübeck

Mecklenburg

Oldenberg

Braunschweig

Anhalt

Schaumburg Lippe

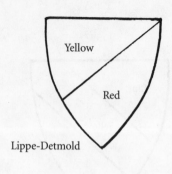

Lippe-Detmold

Yellow

Red

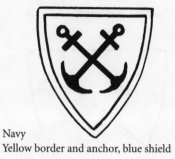

Navy
Yellow border and anchor, blue shield

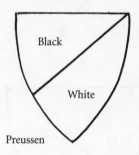

Preussen

Black

White

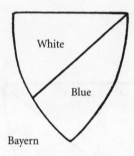

Bayern

White

Blue

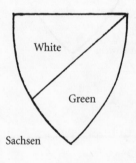

Sachsen

White

Green

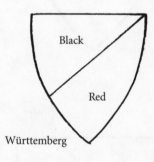

Württemberg

Black

Red

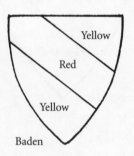

Baden

Yellow

Red

Yellow

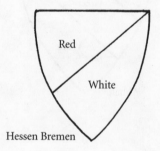

Hessen Bremen

Red

White

44

Three examples of helmets from World War I; (*from left*) special cutout, Turkish visorless, and standard M-16/18. See Fieldmarshal von Hindenberg in the background.

At left: A WWI infantryman. *Above:* Austrian border guards in 1938.

A transitional helmet with Wehrmacht insignia.

THE NEW M-31 LINER

In 1931, a new liner was produced for the Stahlhelm. It was a great improvement over the old liner used in WWI. It was so well accepted that it was used in many different models throughout WWII and was still in use with German and Austrian helmets until the late 1980s.

There were a few changes that set it apart from its predecessor. An expanded selection of sizes was available, thus fitting the head in a more comfortable manner. The mattress system was discontinued, and a new liner subsequently consisted of eight leather "fingers" with holes in each for ventilation. Next, a spring system contained in the liner band provided a bounce and helped ease shock. In addition, the use of aluminum components in place of steel reduced unnecessary weight and bulkiness. And finally, a buckle-type chin strap hooked directly to a swivel connection attached to the liner band, thus making the strap more flexible.

The M-31 liners replaced the old type immediately. They were used with Nazi helmets beginning as early as when Hitler took power. This was what constituted the name "transitional"—a combination of the old helmet and the new liner. It was used later in the M-35, M-35/40, and M-42 helmets.

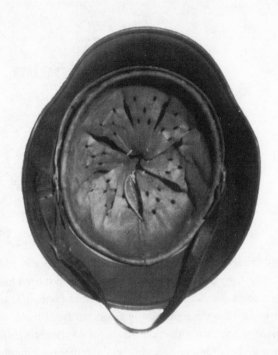

Liner Detail of the M-31

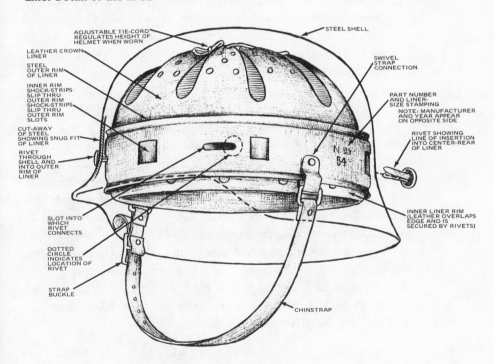

ADJUSTABLE TIE-CORD REGULATES HEIGHT OF HELMET WHEN WORN

STEEL SHELL

LEATHER CROWN LINER

STEEL OUTER RIM OF LINER

SWIVEL STRAP CONNECTION

INNER RIM SHOCK-STRIPS SLIP THRU OUTER RIM SHOCK-STRIPS SLIP THRU OUTER RIM SLOTS

PART NUMBER AND LINER-SIZE STAMPING

NOTE: MANUFACTURER AND YEAR APPEAR ON OPPOSITE SIDE

CUT-AWAY OF STEEL SHOWING SNUG FIT OF LINER

RIVET SHOWING LINE OF INSERTION INTO CENTER-REAR OF LINER

RIVET THROUGH SHELL AND INTO OUTER RIM OF LINER

SLOT INTO WHICH RIVET CONNECTS

INNER LINER RIM (LEATHER OVERLAPS EDGE AND IS SECURED BY RIVETS)

DOTTED CIRCLE INDICATES LOCATION OF RIVET

STRAP BUCKLE

CHINSTRAP

47

SEVEN

A New Stahlhelm
for a New Era
The M-35

As THE NATURE OF war changed, so too did the armies and so their needs. The new German army under Hitler was mobilized and started to become one of the greatest combat machines on the face of the earth. The new military vehicles and weapons were significantly advanced from those of WWI and could be moved much faster. This meant that the infantryman had to move faster. His gear and equipment had to meet contemporary standards of high mobility and firepower.

The designers of the new helmet knew that the upcoming war would not find armies stuck in trenches, and therefore the German soldier would not use the frontal plate any longer. They knew that the new helmet certainly had to pass rigorous ballistic tests, yet it had to be simpler in design for the swift movement that the new mobile units would employ.

A NAZI MODEL OF THE STAHLHELM: THE M-35

The Model 35 which first appeared and was issued to the troops in 1935, was a definite improvement over the M-16, M-17/18, and transitional helmets. The dome, neck guard, and visor were all shortened, resulting in a smaller helmet that provided sufficient protection. It incorporated the new M-31 liner that kept the visor even with the eyebrow. Bushing vents on each side provided for proper ventilation rather than the old side lugs. No frontal plate was anticipated. Unlike earlier models, the M-35 had no shield lugs and no lugs for chin strap. There was a new one-piece liner along with a chin strap held to the liner by a flexible "D" ring, and the bowl of the helmet was a little less deep.

As in the older models, the M-35 shells came in different sizes. The size was stamped in the back on the inner side of the neck guard. The following sizes of helmet are for the listed head sizes:

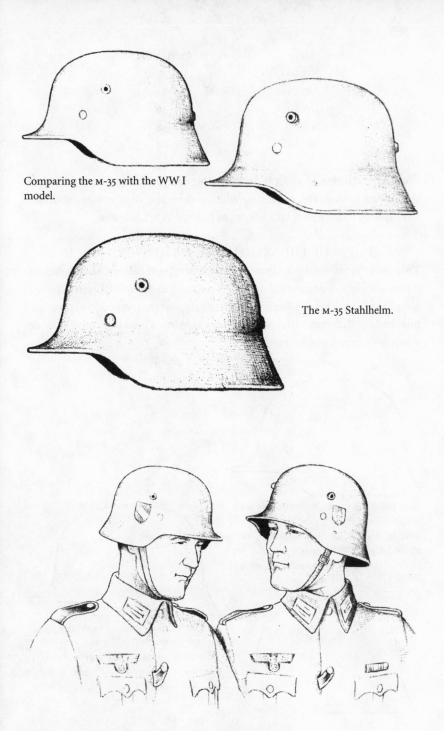

Comparing the M-35 with the WW I model.

The M-35 Stahlhelm.

Two Nazis wearing the M-35 and transitional Stahlhelms.

SIZE	FIT IN CENTIMETERS
No. 60	50–52
No. 62	52.5–54
No. 64	54.5–56
No. 66	56.5–58
No. 68	58.5–60

The manufacturers' identification was stamped in code along with the helmet size on the neck guard, as on the older models. The alloy was molybdenum steel and included nickel and silicon, as had the WWI helmets.

SHAPES OF THE STAHLHELM IN THE THIRD REICH

There were many variations in each particular type of helmet. Also, there were different colors used, and the texture was changed for special purposes. The use of the large selection of early and late insignias that were added presented hundreds of different variants of German helmets. There were nine basic designs, with numerous subtypes.

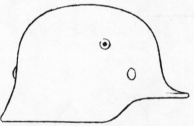

M-35 and the M-35/40. The completely new Stahlhelm was made for the newer type of warfare. It was stamped, of one piece, and had a rolled edge. In 1940 the vent bushing was replaced by a single embossed stamped hole.

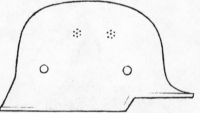

Police. It still maintained Germanic style, except for the square-dip along the side. It had different vent holes, liner, and rivet positions and was usually lighter in weight.

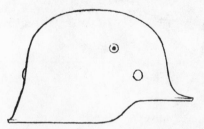

Transitional. From the WWI mold, it was equipped with the new M-31 liner and strap, correct paint, and insignias.

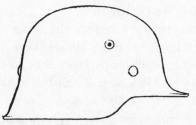

M-42. A revised idea from the earlier model. Almost identical, except the edge was unrolled and more prominent, and the angle of the sides was wider. It was also sometimes labeled the M-43.

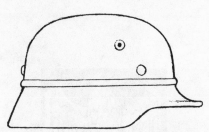

Luftschutz combat style. It was also used by the Luftwaffe. The standard design had a bead addition running completely around the base of the dome.

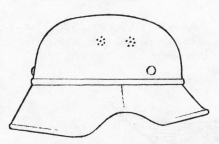

Luftschutz early model. The dome was rounder and had a bead feature and the visor and neck guard were larger. It also had a different liner and vents. A subsequent variation featured a one-piece visor-neckguard.

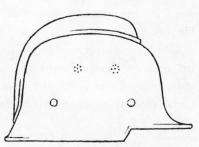

Fireman. Same as the police model but had the addition of a comb, which was either aluminum or chrome.

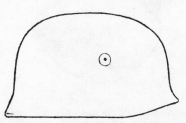

Paratrooper. This Stahlhelm was without a neckguard and visor and with a different liner (heavily padded) and strap. The rivets were unique and were in different locations.

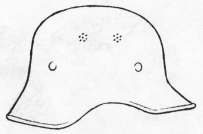

Luftschutz late model. Similar in design to the early model, it had no bead and the liner and rivet locations were different. It was usually lightweight.

At no other time were military insignias displayed and used in such vast numbers and varieties as they were during Hitler's Third Reich. The steel helmet was a fine example because they were all colorfully marked with individual identification in the form of decals. They were all located directly below the side vent or in that general area. It was almost inevitable that in serious combat, those decals would provide attractive targets for enemy snipers at least once. Eventually, most army units were ordered to remove their decals of national colors. However, some individual units did not.

National colors. Germany began using this red, white, and black shield in 1933 after Hitler abolished the old state colors. This was used on the right side of Wehrmacht (army), Kriegsmarine (navy), and nonmilitary helmets.

Wehrmacht. The black shield with a silver-white eagle was used on the left side of the army helmets.

Luftwaffe. The air force helmet displayed the silver-white and black flying eagle on the left side. There were several variations of this design.

State colors. This red shield with its white center and black swastika was on the right side of police helmets and on the left side of SS helmets.

SS. This silver-white shield with two black runes (lightning bolts) signified the SS, with 30 percent of the SS helmets also having the state colors on the opposite side.

Kriegsmarine. The navy insignia was the exact design as the Wehrmacht, except the eagle was gold in color.

Luftschutz. The air defense services began using the silver and black emblem on the front of their helmets in 1938.

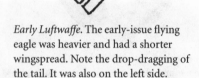

Early Luftwaffe. The early-issue flying eagle was heavier and had a shorter wingspread. Note the drop-dragging of the tail. It was also on the left side.

Polizei. A black shield and a silver eagle and wreath made up the police decal. It was worn on the left side of the helmet, with the state colors on the right side.

Wasserschutzpolizei (not shown). The water police decal was exactly like that of the police, except the eagle was gold in color.

Bahnschutz. A black shield with a blue wheel and gold wings was the decal worn on the left side of the railway police helmet. About 50 percent of the helmets had the national colors on the other side.

Red Cross. A white shield with a black eagle, white swastika, and red cross was the decal on the general issue helmet.

Red Cross. The second example was a white shield with a red cross and white swastika.

Political. This featured a single stylized eagle on a black shield.

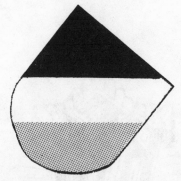

Early police. "Slanted" national colors—red, white, black—were worn on the left side of the helmet.

Early police. A large slanted white swastika was worn on the right side, with slanted national colors on the left.

SA. An eagle appeared on the left side of the helmet. The colors were black and white. The white was sometimes replaced by gold.

SA. A red (maroon) circle trimmed in white, and a white stylized "SA" emblem appeared on the right side of the helmet. The white was sometimes replaced by gold.

RLB. The early air defense helmet's decal, placed in the front, was silver-trimmed in black with blue letters and a swastika.

Reichsarbeitsdienst (RAD). This large black shield
featured a white eagle and wreath.

Afrika Korps. All insignias on the desert
helmets were painted on. No decals were ever
authorized, and therefore there may be
variations in each. Most often, the palm tree
and swastika were white, infrequently on a
green panel background. The crest outline
shape around the tree was also not common.

EIGHT

Fallschirmjäger

The M-37 and M-38 Paratrooper Helmet

FALLSCHIRMJÄGER IN GERMAN is actually made up of three words: "fall," "umbrella," and "hunter." Combined they can be interpreted as "fighter from the skies." The fallschirmjäger, or paratrooper, wore a different uniform and insignia than did the regular ground troops. Also, his helmet was one of unusual design. It is apparent that it was derived from the standard M-35 helmet with some parts omitted or removed. The visor and ear guards were cut to eliminate the protrusive parts for air maneuvering and to enable the headgear to be more compact.

A different strap assembly was provided to prevent the loss of the helmet in the air and also to give more safety to the chin and neck regions from jolts and pressures. What was known as a "chin harness" came in four variations. First was the early issue, which was made of grey leather and was backed by chamois on the side that touched the face. It had a quick-release buckle of two positions with snaps. The developmental issue that came next was wider but was made of a thinner material. Specifically, grey leather with ersatz chamois backing and a standard slide buckle. Then came the late issue which was made of very heavy leather, and the slide straps were riveted on rather than sewn. The harness was much thicker than the first two. Finally, after the enormously costly invasion of Crete, virtually all paratroopers were taken off jump status and used as assault infantrymen for the duration of the war. Their helmets were refitted with a D-ring and a standard buckle like the regular German army helmet. These remade harnesses were not as heavy or bulky as the jump issue.

For shock, more padding was necessary, and special allowances were used in the dome with the support of a rubber lining that ran almost the extent of the interior. This was covered with a leather lining and contained twelve one-inch holes for ventilation.

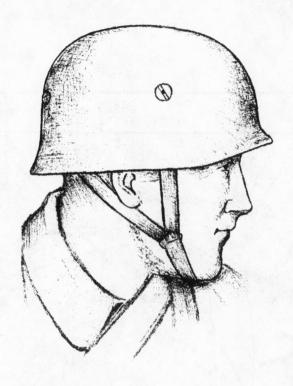

Instead of the regular type rivets for securing the liner to the shell, a bolt and nut system was used. Again, this was for prevention of breakdown that could occur in the air. There are some known examples of raw-edged helmets being produced, but otherwise all paratrooper helmets had a rolled edge. Some examples were manufactured before and at the beginning of the war with Wehrmacht and national colors on the left and right sides. That was before air force commander Hermann Göring appropriated the paratroops for the air force, and those examples are extremely rare. Most paratrooper helmets were decorated only with the Luftwaffe wings on the left.

Helmet covers were not at first supplied, but eventually there were many different types. For example, there were cloth helmet covers that were blue-grey with slots for camouflaging. These were first used on Crete, and others were developed as the troops were used as ground assault forces. Camouflage covers came in two different patterns: an early and late issue. These were the same pattern as the smocks (jumpsuits) held on by clips with slots. Chicken wire covers were made of heavier steel than normal chicken wire. They were sized to the helmet and held on by two clips, one on each side. Net type covers had large holes in the netting. The liner was removed, the net was put over the helmet,

Paratrooper Interior Detail

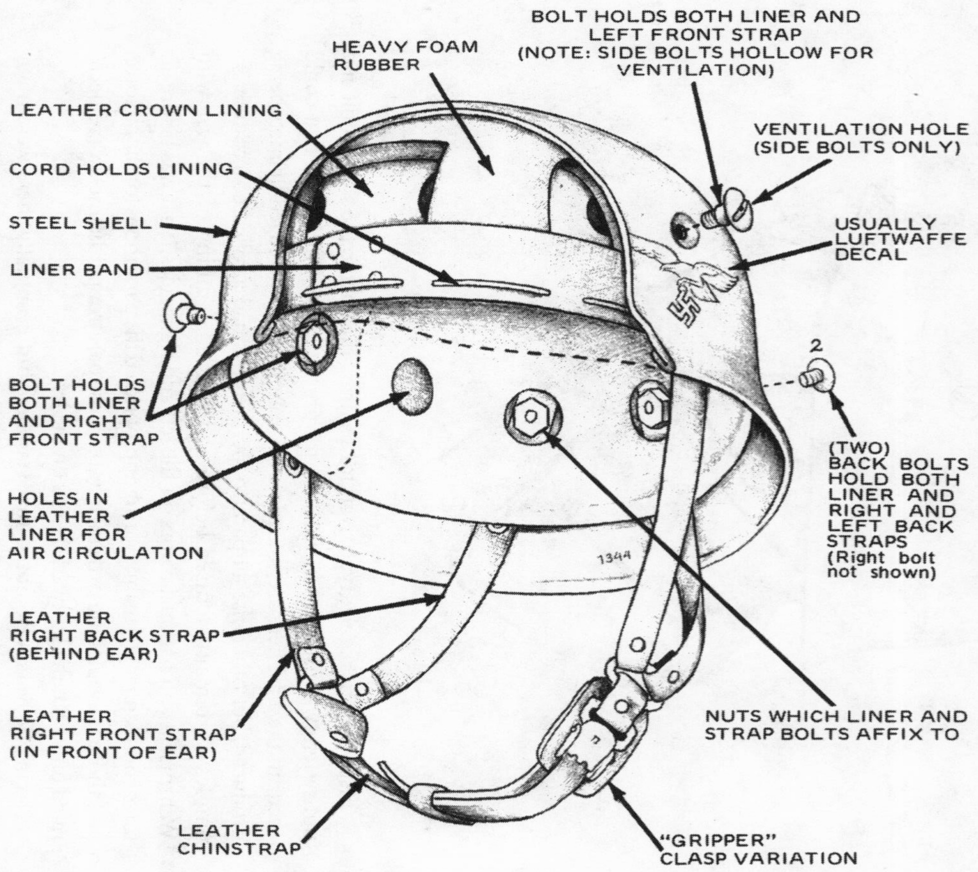

BOLT HOLDS BOTH LINER AND
LEFT FRONT STRAP
(NOTE: SIDE BOLTS HOLLOW FOR
VENTILATION)

VENTILATION HOLE
(SIDE BOLTS ONLY)

USUALLY
LUFTWAFFE
DECAL

2

(TWO) BACK BOLTS
HOLD BOTH
LINER AND
RIGHT AND
LEFT BACK
STRAPS
(Right bolt
not shown)

NUTS WHICH LINER AND
STRAP BOLTS AFFIX TO

"GRIPPER"
CLASP VARIATION

1344

HEAVY FOAM
RUBBER

LEATHER CROWN LINING

CORD HOLDS LINING

STEEL SHELL

LINER BAND

BOLT HOLDS
BOTH LINER
AND RIGHT
FRONT STRAP

HOLES IN
LEATHER
LINER FOR
AIR CIRCULATION

LEATHER
RIGHT BACK STRAP
(BEHIND EAR)

LEATHER
RIGHT FRONT STRAP
(IN FRONT OF EAR)

LEATHER
CHINSTRAP

and the liner was replaced. Also, helmets were covered with mud and allowed to dry. This gave the blue-grey helmet an earthlike color similar to the locale.

The Second Fallschirmjäger Division that saw service in North Africa used the standard paratrooper helmet with Luftwaffe decal insignia. All were painted an Africa Korps tan by the parachute riggers and personal equipment men of each regiment. They brushed on two or three coats of standard paint that was used on trucks and other equipment. While doing this they often covered up the flying eagle insignia. The paint jobs ranged from excellent to poor, depending on how rushed they were and the craftsmanship of the artist.

M-37 HEAD SIZES

SIZE	FIT IN CENTIMETERS
No. 66	52, 53, 54
No. 68	55, 56, 57
No. 70	58, 59, 60

M-38 HEAD SIZES

SIZE	FIT IN CENTIMETERS
No. 66	53, 54, 55
No. 68	56, 57, 58
No. 71	59, 60, 61

NINE

Other Wartime Military Helmets

THE M-35/40: VENT HOLE CHANGE

There was one change in the M-35 Nazi army helmet that should be emphasized. There were different types of vent bushings even within the M-35 series, but the M-35/40 presented a whole new approach. At a quick glance, both models appear the same, but there is an obvious difference when examined closely.

The M-35 models had the vents fitted with bushings. Wartime production helmets omitted the bushings and instead punched embossed vent holes, thus substantially simplifying the process. This latter design is more properly designated as the M-35/40, but it is often referred to as the M-40. This new version did result in an important simplification of production. The 1935 helmet required fourteen separate procedures. The M-35/40 version needed eleven. The M-35/40 also used a new steel alloy based on manganese and silicon.

THE M-42

By 1942, Germany was at the peak of an all-out war. All preparation at home was geared for a long, hard battle, and everyone was preparing for unpredictable times ahead. There was a heavy drive toward sacrificing unnecessary over-designing and avoiding waste. Cautious measures were made to utilize all material for the war effort. Captured enemy helmets were repainted, refitted, and used by organizations such as the Luftschutz and police.

Changes in the quality of the new war implements and garments such as uniforms were recognized at this time. The helmet was no exception. A new, modified Stahlhelm was designed to be produced quicker and cheaper. It was called the M-42.

The M-42 was very much the same in design as the M-35, except that it had no rolled edges. The profile was very similar, though the forward view featured a somewhat more open cross-section. The edge was "raw" and flared outward

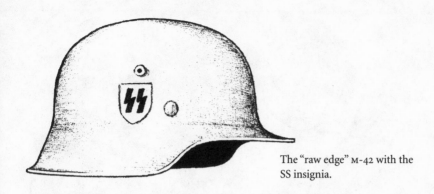

The "raw edge" M-42 with the SS insignia.

slightly for drainage. Some models were fitted with cheaper liners. Many were issued with no insignia decals for the same budgetary reasons. And where the M-35/40 had used eleven manufacturing procedures, the M-42 needed only four.

The helmets were used by all military branches including the SS and were produced until the end of the war. Though manufactured under the conditions of total war, the M-42 provided no less effective protection.

CAMOUFLAGING TECHNIQUES

In many cases, German helmets were repainted with colors to match the terrain and weather so as to avoid detection by the enemy. For example, white was added to match snow and tan was used for desert fighting. Sometimes dirt and gravel were mixed with the paint for a more nonreflective texture. This

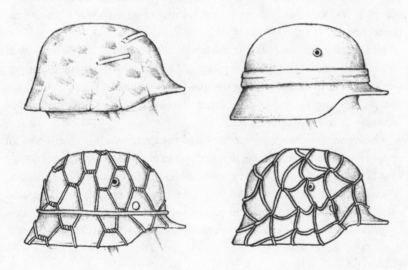

painting process was usually performed by the individual helmet owners. There-fore, there was no standard camouflage pattern. In some known cases, the paint was simply sprinkled or splashed on with crude instruments like bits of cloth-ing, rocks, and sticks.

Camouflage covers were also issued to wrap about the helmet. Their design enabled mud, leaves, branches, and other material to be attached. They included cloth covers (many with special seasonal printed patterns), chicken wire covers, cloth straps, and netting. The variety of cloth covers increased as the war pro-gressed. Different branches of the military developed unique patterns.

THE JET HELMET

The jet helmet cannot be overlooked in the history of Nazi combat helmets because it belonged to the family of "steels." Not nearly as numerous as the others, it was valuable and important because in its own way it was a true pioneer of the jet age.

Jet aircraft made their entry in the final battles of WWII, and Germany was undoubtedly their most effective pioneer. Another newcomer late in the war was the ejection seat, which was designed safely to thrust the pilot from the cockpit and over the tail of a superfast disabled aircraft in a matter of split seconds—in a sense, automatically.

These two new developments required the design and production of an appropriate jet helmet. The ejection seat was in more of an experimental stage in 1944 to 1945 compared to the proven models in the world today. Therefore, it was not unlikely that the pilot would have had to eject through a closed canopy that was stuck. Thus, the pilot was put in danger by numerous new hazards. This pervasive risk was the reason for using steel inside the helmet. A

padded comb extended around the top and down the front, which acted as a bumper and also provided extra protection from impact and shock.

The steel layers were covered with good quality brown leather. The inside was lined with a soft chamois-type cloth. The strap was a double unit and attached to the helmet by the use of snaps. It was also made of the same high quality and color of leather, as was the cover for the rubber comb. A cut-out was located at the ear area to allow the use of earphones.

Very few of the Luftwaffe pilots who maneuvered those early jets were issued the jet helmet, let alone had knowledge of its existence, since it was developed at the very end of the war. Nonetheless, it was a pioneering effort that had critical implications for the future of jet flight.

TEN

Nonmilitary Helmets

LUFTSCHUTZ HELMETS

Several types of steel helmets were used for nonmilitary purposes during World War II. Some of these were the ordinary line of combat helmets and were distinguished only by a particular paint color and the insignia of the specific organization that used it. However, there were two different body styles designed exclusively for non-military purposes that were issued to nonmilitary organizations and most widely used by the air defense service (Luftschutz). The two were commonly known as "Luftschutz helmets," although many were used for other purposes. Specifically, the two were called the "gladiator" and the "combat" types.

The early gladiator Luftschutz was made up of three separate pieces welded together. It was known as the M-38. The liner was a much cheaper product than the regular M-31. The second gladiator Luftschutz resembled the early model, but it was a simplified two-piece helmet known as the M-39. The liner was also somewhat crudely made. The third type of gladiator Luftschutz helmet was a kind of last-ditch affair stamped and drawn as one piece. It was simple by comparison to the earlier models, but it still afforded reasonable protection during an air raid. This late model was termed the M-44. A flat rimmed variety was made in small numbers.

The combat Luftschutz was a lighter version of the same basic helmet design profile as the army style M-35 except for a bead running around the crown. This housed the same type of liners as found in any of the other Luftschutz models. It should be noted that not all beaded combat models were Luftschutz. Some were simply regular full-weight helmets with some kind of defect. The bead excluded it from being issued to regular forces.

All four Luftschutz helmets were lighter than the ordinary combat type helmet. There were selections of colors involved, but until very late in the war a dark blue was used for air defense. Air vents differed, but usually they were of

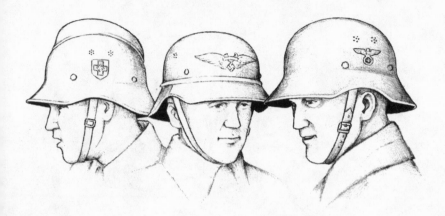

the same type as on police helmets: seven small holes grouped in a round half-inch area. And, often, two vent-hole groups were found on each side. All had wide chin straps with matching colors.

Almost all of the Luftschutz helmets had the special Luftschutz decal insignia on them. It was a large, silver wing with swastika and Luftschutz identification. The manufacturers' marking and helmet sizes were stamped on the back of the inner section of the neck guard.

Red Cross workers and firemen used such helmets, as well as special police and combat styles, but the helmets were painted another color and almost always carried the insignia or decal of their organization. A comb was added in many cases for extra significance. These were either made of painted metal, chrome, or aluminum and they were connected through the top of the helmet by small bolts, with a felt washer between to seal off any cracks.

The gladiators were made in ten sizes, but there was no standard relationship between the helmet and the fairly simple liners. Thus, no shell-head table can be usefully constructed.

POLICE AND FIREMAN HELMETS

During WWII civilian police units were equipped with steel helmets with a Germanic design. Each type was unique in this noteworthy array of helmet shapes. The steel was thinner than the usual M-35 combat helmet and therefore considerably lighter in weight. However, sometimes there was a wide variation of steel thickness and the weight of the helmet that made it similar to the ordinary Stahlhelm.

A unique feature of the police models was that they had a profile featuring square dips in their ear guard section rather than the regular slight curve. The ordinary ear guard style of 1935 was one of the few exceptions. Most models

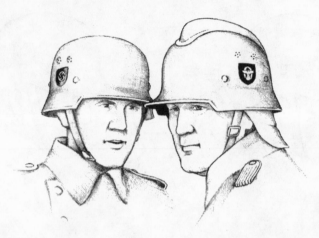

had two sets of vent patterns made up of seven holes in a half-inch circle on each side of the bowl. There was also a type of police helmet that had insert cups with holes for the inserts.

The liners were made of a smooth leather and came both with and without the spring band. The straps were usually wider than the standard military model and many were secured with the Y style chin strap.

The early police helmets' insignias were different than the later insignias in a couple of ways. First, the earlier model had a large, three-inch white swastika (slanted), on the right side and the national colors (slanted) on the left. And second, the later issue had the silver police insignia on a black shield on the left side and usually the state emblem on the right side. All police helmets were black with few exceptions.

The fireman's helmet was exactly the same as the police model with the exception of the addition of the comb on top. No matter what style (other than the typical police model) the fireman used, it could be recognized from great distances because of the unique aluminum comb.

The liners were the same as on police helmets. Sometimes leather neck protectors were added that were attached to the liner with the help of five leather lugs. The leather neck protector could be stuffed up into the liner, however uncomfortably, and the helmet could also be worn without it. When released, it hung down the side and back portion of the fireman's neck and offered substantial protection.

ELEVEN
Germany and World War II

IN MARCH 1938, GERMANY absorbed Austria by simply marching in unopposed, while the domination of Czechoslovakia came in a slightly more complicated manner. The population of "greater Germany" was increased by over ten million people. On September 1, 1939, Germany invaded Poland assuming that the world would accept that outrage as well. On the third of September 1939, however, Britain and France declared war on Germany, and World War II began.

As previously mentioned, the period between September 1939 and April 1940 was known as the phony war, or *sitzkrieg*, since none of the three combatant nations made a major move. The Russians, however, were not idle. On September 17, 1939, Russia, through their agreement with Germany, invaded Poland from the east and divided it up. As 1939 gave way to 1940, Russian troops attacked Finland, which surprisingly held out for three months. They also occupied the Baltic states and the part of Romania known as Bessarabia.

In April 1940, Germany swiftly conquered Norway and Denmark. Then on May 10, 1940, the Germans changed the sitzkrieg to the blitzkrieg and blasted into Holland, Luxembourg, and Belgium and then on into France through the Ardennes Forest. France surrendered in June 1940. The French army and British Expeditionary Force suffered a crushing defeat that climaxed at Dunkirk when a makeshift British flotilla evacuated thousands of French and British troops to safety.

Germany had a firm grip on "fortress Europe." The Battle of Britain ensued in which the German air force attempted to defeat Britain's resistance through daylight air raids. However, the Royal Air Force, together with the newly developed radar and the British ability to read German ciphers, combined to neutralize the Luftwaffe. German air losses were so great that daylight bombing was abandoned in favor of frightening, but not very effective, night attacks. When heavy losses continued and England failed to crack, the Battle of Britain was over and the British had prevented invasion of England: There could be no attempt at invasion without air superiority over the Channel. At the time, it appeared that England's defeat was only a matter of waiting for the right moment.

With England militarily immobilized and no longer a threat in the west, Germany turned its attention to the east and Russia. In October 1940, Italy invaded Greece. The Italian army was stopped, then pushed back to the mountains of Albania. The British landed over fifty thousand troops in Greece, occupied Lemnos and Crete, and threatened the back door of Hitler's Europe. The British had also kept a firm grip on the Suez Canal lifeline and what they called the western desert. The greatly outnumbered British in North Africa, on the verge of defeat by Christmas of 1940, began a counterattack with three divisions that resulted in the annihilation of ten Italian divisions by mid-February 1941. On April 6, 1941, in Libya, Germany again demonstrated the blitzkrieg, that fearsome combination of advance heavy bombing, dive bombing, concentrated tank units, and motorized troop carriers. Paratroops destroyed or secured objectives miles ahead of the advancing armies. The greatly improved electronic communications allowed for immediate coordination of air and ground forces.

By the end of May 1941, the Germans had regained the ground lost by the Italians, had defeated Greece, and had taken the island of Crete. Crete was conquered exclusively by air as the Luftwaffe chased off the British fleet in the waters around the island while an airborne force dropped their might from the sky. The British troops in Greece found themselves facing another Dunkirk, and many were evacuated to North Africa. In late February 1941, two months before the German attack on Greece, the Germans had landed three divisions in North Africa, one German Panzer and two Italian, one of which was armored. The British, having defeated an overwhelming Italian force just a few weeks before and itself reinforced, confidently awaited the enemy's appearance. Again the blitzkrieg struck, and within three weeks the British were facing the loss of all of North Africa, including the Suez Canal.

British reinforcements and supplies brought the North African war to a near stalemate as the German Afrika Korps was neglected by the German General Staff, which had more important things on its mind: the invasion of Russia. Supplies and reinforcements scheduled for the Afrika Korps never got there because the full German war effort was concentrated on preparations for "Operation Barbarossa."

On June 22, 1941, Germany invaded Russia. The blitzkrieg was never more devastating. It looked as though the war would be over within two weeks, with all German military expectations easily surpassed. (An interesting footnote to the history of the war in the Soviet Union was that in desperation, the Russians opened their warehouses full of German and Austrian helmets captured during WWI. The army issued 1916 and 1917/18 models, painted with large red stars over the original finish, to whole combat formations to be used against their former owners.) On December 2, 1941, a German reconnaissance outfit was close enough to Moscow to claim to have seen the Kremlin domes. While that would have been physically impossible, they got very close. Russian military strength and capability had been hopelessly underestimated by German military intelligence. That factor plus the lend-lease equipment sent to Russia by the United States and Britain finally began to assert itself. On December 2, 1941 the German army approaching Moscow was driven back in a massive counter-attack that all but completely routed the entire German Army Group Center.

Meanwhile, the Afrika Korps, which was rapidly running short of supplies and manpower, watched the Suez Canal traffic increase as British strength grew in a greater ratio than Afrika Korps losses. The tide had turned.

On December 7, 1941, Japan attacked the United States at Pearl Harbor. With that attack, Germany's last hopes for victory disappeared. They had long hoped Japan would invade Russia from the east and relieve the pressure on German armies west of Moscow. However, Japan honored its nonaggression treaty with Russia and pursued their military objectives in Southeast Asia and the Pacific. With the United States actively in the war, supplies began to pour into England, North Africa, and Russia, and the German military began to stagger under the increased force of counterattack. The Russians could not immediately follow up their success and were bogged down in the spring mud of 1942. Both sides paused to regroup. The German army alone counted its losses in the first nine months of the war in Russia as over one million casualties.

In North Africa, the Afrika Korps, which was on the verge of collapse by December of 1941, was belatedly reinforced with three German and eight Italian divisions. In January 1942, the Afrika Korps won a series of victories in a seventeen-day period that completely changed the North African campaign situation again.

At first the German army in Russia was bolstered by the aid of its allies' armies. Romanian, Hungarian, Italian, and Slovak armies had deployed forces to the east, along with Finnish troops who sought to take the opportunity to redress the losses of the Winter War. Spectacular successes were achieved initially by the twin drives in Russia and North Africa. By mid-August 1942 the German army had reached Stalingrad in Russia, captured the Maikop oil fields, and were within fifty miles of Grozny, a main Russian oil refining center, and one hundred miles from the Caspian Sea. If Stalingrad fell, the Russian armies would be cut off from a major supply of petroleum products that were crucial to the war effort.

The Afrika Korps also attacked in May 1942, driving into Egypt threatening British control of the Suez Canal. They occupied Tobruk, Britain's military center in North Africa, without strain. The Afrika Korps was at El Alamein with the Nile just beyond, and Stalingrad was under heavy attack in Russia. A giant pincer was closing whereby the German Afrika Korps, moving northeast through the oil-rich Middle East, would link up with the German armies in the Caucasus. It was at this point that the overextended German war machine ran out of gas and ran into ferocious counterattacks. Dwindling supplies, lack of equipment and manpower reserves, together with the influx of war material from the West, swung the offensive to the Soviets, and the Germans began fighting a retreating, defensive war.

After the German offensive failed at El Alamein, the British began a push forward. By mid-November 1942 the British had recovered over seven hundred miles of North Africa for the Allies, and the battered Afrika Korps continued to retreat in the face of ever-increasing odds. In November 1942, the Allies landed in Morocco and Algeria, and the Afrika Korps was caught between the two forces. In May 1943, the remnants of the Afrika Korps surrendered, and the war in North Africa was over.

The Germans in Russia had split their forces in order to take Stalingrad and the Grozny oil fields at the same time. In full force, either objective may have been accomplished by the Germans, but divided it was impossible. Fighting in Stalingrad raged from building to building. Russian reinforcements broke through the German and allied flanks and 250,000 Germans were cut off. Hitler refused to allow any attempt to break out and all rescue attempts failed. Twenty-two divisions were lost to the German army as well as Stalingrad. The armies at Grozny were ordered to retreat in the face of a second "Stalingrad" as the Soviets—employing millions of men, and thousands of armored units and having largely gained control of the skies—threatened another mass encirclement.

The entire German military force in Russia was in sporadic but inevitable retreat and was in periodic danger of being cut off by the advancing Soviet army. Surrounded with nowhere to go, the remaining forces of the Sixth Army at Stalingrad were given a twenty-four-hour ultimatum to surrender on January 8, 1943, which they rejected. They held out almost a month longer, but on February 2, 1943, the last shot was fired and Stalingrad was still. It was the worst defeat in all of German military history. And the war was now in Germany, closing in as thousand-plane bomber raids by 1943 were hitting all of its major cities with greater frequency and were meeting less resistance. Allied forces were attacking at will, day and night.

Within two months of the conquest of North Africa, the Allies invaded Sicily in July 1943. At that same time, the Germans launched an armored attack at Kursk featuring new and heavier tanks and assault vehicles. Allied signals intelligence had provided the Soviets with the complete plans. Within a few weeks, the drive stalled and the Soviets began their summer offensive, moving the Germans back along the entire front.

On September 3, 1943, the Allies invaded the Italian mainland. The Italian military leadership signed an armistice with the Allies after a successful coup overthrew the Mussolini government. However, a countercoup organized by the Germans restored the fascist government in the unoccupied areas. The Allies landed south of Naples a week later, and the hard-pressed German forces in Russia were brought in, further weakening the Eastern Front.

The Germans fared better in Italy. South of Rome their defenses managed to hold the Allied advance. But by that time, the Allies had complete superiority in the air and on the sea. New radar had driven the submarines out of the Atlantic, and the convoys hauling massive cargoes of supplies needed to launch an invasion of fortress Europe itself managed to get through from North America to England intact.

The massive strategic bombing raids on German cities created a new problem of modern warfare: millions of homeless, starving civilians. Soldiers worrying about their wives, children, and parents back home were, in many cases, more psychologically affected than the civilians. The winter of 1943–44 was a bitter time for German civilians and military alike. In January the Russians were on the verge of taking the Romanian oil fields of Ploesti, the principal remaining source of petroleum products for the German armed forces.

The spring mud of 1944 and Stalin's erratic wartime strategy slowed the war in the east again, and the huge buildup of supplies and troops in England caused the Germans to shift forces from the east to the west anticipating an Allied invasion of the Continent across the English Channel. Many of the troops from the east were, in fact, former Soviet soldiers who had chosen to change sides in order to fight Stalin. In June 1944 they found themselves in the unlikely situation of fighting Churchill and Roosevelt. On June 6, 1944, the Allies landed in Normandy. It was the largest amphibious invasion force ever assembled. The coast at Normandy was bombarded from the sea and the air, and the landing was successful. The Allies could now begin the liberation of occupied Western Europe.

On June 10, 1944, the Soviet army began its final summer offensive against Germany. By September 1944 the Russians had captured the oil fields in Romania, Bulgaria had changed sides, and the Finnish troops had surrendered. In the west, the German armies were in retreat, and Belgium was in Allied hands. In mid-September 1944 the Allies had outrun their supplies and were bogged down west of the Rhine. The Soviets had halted their offensive in Poland and East Prussia for strategic postwar motives.

The Germans, realizing they could not just sit still and wait, launched a counteroffensive in the west on December 16, 1944, driving a narrow corridor through to the Meuse River. However, exactly one month later, the Battle of the Bulge, as it became known, was over and the Germans were back where they had started. On January 27, 1945, the Soviet army pushed a winter offensive and was only one hundred miles from Berlin. By March 1945, virtually all German mechanized weapons, including naval vessels, were idle due to a lack of fuel and ammunition. The v-1 and v-2 rockets that had been so threatening to the West were largely lost when the Allies took the launching pads in France, Belgium, and Holland. More than one thousand new jet aircraft were also lost to the Luftwaffe due to lack of fuel and the constant air bombardment of their bases. Of the 126 new super u-boats, only two were ever completed.

In April 1945, the German forces in Italy surrendered, and at about the same time, with Berlin an easy objective, the Allies finally linked up with the Soviets. On April 21, 1945, the Russians reached the edge of Berlin and street fighting began. At midnight on May 8, 1945, Germany surrendered. World War II in

GERMANY AND AXIS

Country	Peak strength	Battle deaths	Country	Peak strength	Battle deaths
Bulgaria	450,000	10,000	Italy	3,750,000	*77,494
Finland	250,000	82,000	Japan	6,095,000	1,219,000
Germany (inc. Austria)	10,200,000	3,500,000	Romania	600,000	300,000
Hungary	350,000	140,000	*Includes 17,494 on Allied side.		

ALLIES AND ASSOCIATED POWERS

Country	Peak strength	Battle deaths	Country	Peak strength	Battle deaths
Australia	680,000	23,365	New Zealand	157,000	10,033
Belgium	650,000	7,760	Norway	45,000	1,000
Canada	780,000	37,476	Poland	1,000,000	320,000
China	5,000,000	12,200,000	So. Africa. Union of	140,000	6,840
Denmark	25,000	3,006	United Kingdom	5,120,000	244,723
France	5,000,000	210,671	United States	12,300,000	291,557
Greece	414,000	273,700	U.S.S.R.	12,500,000	7,500,000
India	2,150,000	24,338	Yugoslavia	500,000	410,000
Netherlands	410,000	6,238			

OTHER POWERS THAT DECLARED WAR ON GERMANY & AXIS *

Country	Peak strength	Country	Peak strength	Country	Peak strength	Country	Peak strength
Albania	25,000	Czechoslovakia	150,000	Honduras	3,500	Peru	40,000
Argentina	160,000	Dom. Republic	5,000	Iran	120,000	Phillippines	200,000
Bolivia	10,000	Ecuador	9,000	Iraq	47,000	Turkey	850,000
Brazil	200,000	Egypt	54,000	Liberia	1,000	Uruguay	11,000
Chile	60,000	El Salvador	3,500	Luxembourg	1,000	Venezuela	15,000
Columbia	19,000	Ethiopa	38,000	Mexico	70,000		
Costa Rica	500	Guatemala	6,000	Nicaragua	3,500		
Cuba	20,000	Haiti	4,000	Paraguay	10,000		

*Forces engaged and losses, if any, not available

NEUTRALS, Peak Strength–Afghanistan, 92,000; Portugal, 115,000; Saudi Arabia, 8,000; Siam, 126,500; Spain, 850,000; Sweden, 350,000; Switzerland, 650,000.

Europe had lasted five years, eight months, and seven days and had ended in unconditional surrender by Nazi Germany.

POSTWAR EAST GERMANY: FROM VOPO TO NVA

Conquered Germany was split down the middle by a demarcation line running north and south. The west side of the line was occupied in zones by the Allies, while the Soviet Union and Poland occupied the east.

Poland occupied the East German districts of Pomerania, East Brandenburg, Silesia, and the major part of East Prussia. The Soviet Union held the East German districts of Brandenburg, Saxony, Saxony-Anhalt, Thuringia, Mecklenburg-Pomerania, and the rest of East Prussia. East Prussia was incorporated into the Russian Republic of the USSR.

In June 1945, within thirty days of Germany's unconditional surrender, the USSR began the formation of a future East German satellite army with the creation of a German police force at Dresden. Within six months, all five German districts in the Soviet zone of occupation had a German paramilitary police force, technically under the control of the zone district's mayor or president. Actually, it was totally controlled by the Soviet authorities. The police forces became known popularly as the People's Police (Volks Polizei) or, for short, the Vopos.

The Soviet military administration that was headquartered outside Berlin at Karlshorst, ordered the formation of another bureaucracy, the German Administration of Interior (GAI). It was placed in the hands of politically dependable German Communists, was carefully overseen by Soviet security personnel, and was given central power over the Vopos of all of the Soviet zone. In 1946, the GAI ordered the police chiefs in the five districts to form a significant new branch of the militarized border police. The new East German frontier police were to be housed in barracks—a distinctly military move—and armed with the available weapons of the vanquished Wehrmacht.

By September 1947, there existed an East German militarized police force of over 50,000 men. Since this paramilitary buildup of East German forces was in violation of the agreement between the Allies, it was rightly assumed that the Communists' use of the term "police" was simply used for camouflage. The uniforms in 1947, including helmets, were either Soviet or of Soviet design. In mid-1948 when the police force had grown to 80,000 or more men, the Russians began an East German military buildup in earnest. The GAI ordered another new arm, the Barracked People's Police (BVP), which was to be the official training force for a new German officer corps. Recruits were sought among former Wehrmacht NCOs and junior officers still held in the Soviet Union as prisoners of war. From the BVP, the Soviets were beginning the training of officers for the people's police (army) and the two future military forces, the navy and air force.

At the end of 1948, the Transport Police (Trapo) were formed for the purpose of guarding railroads and the shipments of goods to Russia. At that time there were about 8,000 German officers in training in the BVP under the guidance of Soviet advisers. In the spring of 1949, the Soviets purged the East German military of those they deemed politically unreliable personnel, having exhausted the experience of these mainly nonproletarian veterans. It was essential to the Soviets to keep the West as uninformed as possible about their East German military activities. High-ranking officers of the BVP were being trained in the USSR to ensure the adoption of Soviet military technique as well as political indoctrination.

EAST GERMAN MILITARY:
THE NATIONAL PEOPLE'S ARMY (NVA)

In February 1950, an elite corps of 4,000 trusted Communists was set up as the Guard Regiments of the Ministry of State Security. At first these included a first and second army regiment. The first and second flotillas of the naval police were then created, and early in 1951, with the subsequent formation of the air police, the East German military thus consisted of all three branches of land, sea, and air forces. As separate units they stood ready for individual expansion.

In March 1952 the USSR began to stress the importance of military duty to the East Germans in order to combat the aggressors in the West and suppress the forces of reaction at home. They urged service in the East German paramilitary forces, a three-year voluntary enlistment, as a duty and privilege. However, as a combat force, the East Germans were still too weak and poorly trained and equipped to be an effective military formation.

Combat groups were formed, as a "home guard" or workers' militia force as had been done in other Soviet-bloc countries. Their ranks were boosted by returning discharged police soldiers who had finished their three-year enlistments. The Soviets were now able to work at an East German military buildup with a standing army of reserves trained in the Russian military style under careful watch.

In 1953, East German forces consisted of seven divisions that were divided into Army Groups North and South. In 1954 the three paramilitary branches fell under the Supreme Command of the Deputy Chairman of the Council of Ministers, a German Communist, who was supervised directly by Moscow.

By 1955, the former Wehrmacht officers, who were considered a necessary evil because they were valued only for their military experience, had given way to younger, more politically reliable East Germans trained by Soviets. In October 1955, the first massive field maneuvers were held, and by the dawn of 1956, the Soviets had accomplished their East German military objectives: They trained an

army in the Russian style and created a politically reliable German Officer Corps. The three major branches of the military were manifested—air, land, and sea—plus a standing reserve force. Finally, a solid premilitary program for ages fourteen to twenty-four, the Freie Deutsche Jugend or Free German Youth was formed. This program was centered around sports but was paramilitary in nature.

Following the Western decision to create a Bundeswehr in the new West Germany, the Soviets dropped the charade and officially created the East German Armed Forces (NVA). In line with the appeal to East German patriotism and to bolster flagging enlistments, the East German military was given a more Germanic appearance. The traditional German military uniform style was adopted as the standard army issue, but the Soviet-style helmet was retained. One example of a quasi-Germanic helmet had been used by the Vopos, but had been withdrawn in the early 1960s. The old Wehrmacht weapons were phased out and replaced gradually with more recent Soviet equipment. Training was upgraded, and the Soviet advisers faded into the background though they were never far away.

In 1957, the German frontier police force was divided into military units, and the first anti-aircraft batteries were formed. In 1958 two additional armies were formed, and the Soviets began to feel a certain amount of confidence in the East German military, although it would always be said that the East German armed forces were the most trustworthy but the least trusted force among the Warsaw Pact allies. The period from 1960 to 1962 saw an intensified drive to bring the East German army up to date with new Soviet equipment. The Berlin Wall, built by the USSR in 1961, and the intensified confrontation with the West seemed to force the Soviets to overcome their fear of a real East German military as long as it was closely supervised by Soviet personnel.

In January 1962, a General Regulations order was issued enacting national conscription. An oath was administered to draftees whereby they swore unconditional obedience to and support of the Soviet army. In 1963, army, navy, and air police officers were trained at separate centers, and the East German military started to receive Soviet military equipment such as the latest in jet aircraft, ground-to-air missiles, nuclear missiles (with the warheads firmly under Russian control), and other advanced rocketry.

By the end of 1963, the East German military was a major force capable of combat. With its training and Soviet equipment, it was considered a vital front-line link in the Soviet military bloc. It was granted the privilege of serving under a joint Warsaw Pact command and enjoyed perhaps the highest prestige for combat readiness in the Soviet bloc outside of the Soviet army itself. The importance of the East German military as a front-line force was evident through the fact that it was equipped as a very high priority.

The Vopo Stahlhelm

TWELVE

Stahlhelm in East Germany

The M-56 Vopo-NVA Helmet

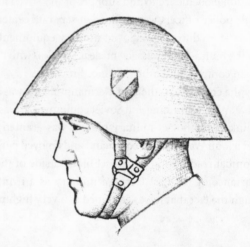

ALTHOUGH THE VOPO HELMET superficially resembled both the Soviet and Germanic pieces, it was a unique pattern. In fact, it was developed, along with several others, in 1942 as a possible replacement for the 1935–40 models. Nothing ever went beyond the initial design effort, but it was waiting when the need for a uniquely new German Vopo-NVA helmet arose.

The old German uniform style was restored to inspire pride and dignity within the East German population, but the Soviets demanded that a new conception be used for the combat headgear. Of course, they would never have allowed the reintroduction of the wartime Stahlhelm. The symbolic impact of the 1935 silhouette was simply too strong for the Soviets to tolerate.

The dome of the 1956 model was rounder on top, and the sides flared out in all directions. The visor and ear guard were then a continuation of this vivid flare. The inside offered padding and warmth and was initially lined completely with cork. There were two types of straps used: a two-piece that buckled with a prong and the Y type, similar to that used in WWII police helmets. The basic design evolved through three specific variants, and a light plastic parade version was also produced. The helmet was sold widely to Third World armies,

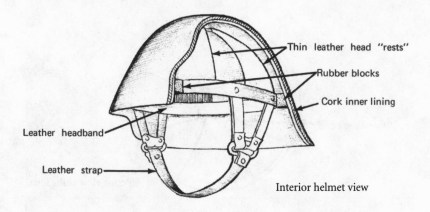

Thin leather head "rests"

Rubber blocks

Cork inner lining

Leather headband

Leather strap

Interior helmet view

and since German reunification it has even shown up in such unlikely spots as Croatia, bought as cut-rate but high-grade surplus. It should also be noted that small numbers of the Polish airborne helmet were purchased and used by the NVA's special parachute forces.

Description
Material	Swedish steel
Gauge	3 mm, (.118 in)
Weight	1 kg (2.205 lbs)
Size	18 cm high (7.086 in)
Color	greenish gray

MARKS AND INSIGNIAS

Police insignias representing various units were used at first. Small GDR shield decals of black, red, and gold on the right sides were also used. And in the cases of real police formations, unit decals were applied to the left sides as well. NVA helmets generally bore no insignias. There were several covers available, the standard being the "rain" pattern. The words "DDR Eigentum" (GDR Property) were stamped on the inside of the helmet.

LINER AND STRAP

The M-56 at first had a hard leather chin strap, approximately one to two centimeters (.393–.787 in) wide and were adjustable by means of a buckle. Most models seemed to have had deluxe "Y" type straps. Leather head rests were available in three general sizes, each adjustable, so that there was no limit to the number of sizes for the helmet. Plastic eventually replaced most of the metal hangers.

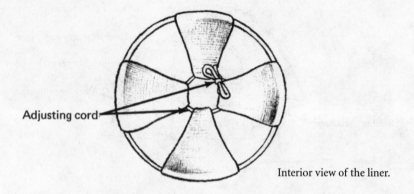

Adjusting cord

Interior view of the liner.

The inside of the earlier version was lined with cork covered with leather, except for two millimeters (.0787 in) of the rim. The cork was soon removed, as it was unnecessarily expensive and of no significant effect in providing warmth.

THIRTEEN

West Germany

The Bundeswehr

THE FIRST FIVE YEARS of the occupation of West Germany by the Western Allies—the United States, Britain, and France—were relatively uneventful in the history of the West German military. For all essential purposes, a military did not exist and the state police forces were not militarized as they had been in the East.

June 1950 was the turning point with the outbreak of war in Korea and the subsequent U.S. military involvement there. It highlighted the growing political and military tensions between East and West. When the Soviet foreign minister drew an analogy between Korea and Germany, the NATO allies recognized that a distinct possibility of war existed in Germany. The question of German rearmament arose. Finally, in 1954, after much bickering within the North Atlantic Treaty Organization (NATO), the United States and Britain pushed through a policy allowing the formation of West German military forces. Called the Bundeswehr, or the federal armed forces, it was initially voluntary. The first German soldier did not don a Bundeswehr uniform until 1955, a full ten years behind his East German paramilitary police counterpart.

In 1956 a conscription act was passed by the West German government calling for compulsory military service for a period of eighteen months. The first

West German soldier was not drafted until, ironically, April Fools' Day of 1957. That year, for budgetary reasons, the United States announced an intended troop reduction from its zone of occupation. In retaliation, the West German government immediately instituted a series of "readjustments." The draft period of service was reduced from eighteen months to twelve months (it was restored to eighteen months in 1962); the ratio of volunteers was increased to 60:40, thereby reducing an effective active reserve; and in order to meet the officer shortage, West Germany lowered the restriction against former junior Waffen SS officers serving in the new army. The target of 400,000 men was reduced to a goal of 325,000. The West German government reasoned that a smaller army with nuclear weapons would constitute a quality army in event of war. Of course, it was unlikely that the Bundeswehr would ever get control over nuclear weapons, but it was left as part of their manning doctrine.

By 1964 the Bundeswehr was beginning to catch up with the NVA. There were twelve or more full divisions organized into brigades of 3,500 to 4,000 men each that were copied from a U.S. airborne style of organization. With five brigades to a division, the Bundeswehr became a highly maneuverable force on the U.S. military pattern. Brigades could be used effectively with or without nuclear weapons.

The West German air force, once again called the Luftwaffe, consisted of sixty squadrons of two thousand airplanes. They included some of the latest U.S. jets and a personnel establishment of 100,000 men.

The new navy floated 200,000 tons, including over two hundred ships, light surface vessels for use as missile launchers, and coastal patrol boats. The new West German navy was allowed a few submarines, but only small ones with very limited range. The new navy put 25,000 men in uniform.

The three separate branches of service were ruled by a joint command. The Federal Defense Minister was the Chief of the Joint Command, and he was responsible directly to the parliament. Thus civilian control was maintained over the growing military, similar in structure to the U.S. military-civilian system.

Two other noteworthy German armed forces were already operational by 1957. They were not like the Bundeswehr, and beyond the scope of NATO. They included a territorial defense force, a national guard–like contingent of 25,000 to 30,000 men with an active reserve of 250,000. Its primary function was engineering, signals, and air defense. In addition, the border security force of 24,000 men deployed along the demarcation line had been organized to provide general border control and to prevent minor intra-German border skirmishes from turning into major incidents.

The Bundeswehr was roughly equal to the East German army in strength and was possibly the strongest military force in Europe other than the Soviet armed forces. One million strong, well-trained, and adapted to the most recent U.S. military tactics, the Bundeswehr was an army to be reckoned with.

When German technical research was given the green light, it signaled a new era of mutual assistance. German scientists and engineers could contribute to the design of new Bundeswehr weapons. By the end of the Cold War, German industry contributed nearly 100 percent of the Bundeswehr's military requirements and had entered in to numerous joint research and development weapons programs with its NATO allies.

STAHLHELM IN WEST GERMANY: THE BUNDESWEHR

There were three basic styles of helmets worn in West Germany prior to reunification. Two of these were designed in the Germanic style and the other one was greatly influenced by the American helmet.

The German M-35/40 style helmet was retained and used by border guards. The liner band was changed, and the shell was lighter metal, but it still maintained all the characteristic appearances of the WWII combat helmet. A plastic parade model was also issued.

The German-American helmet model 1962 was a combination of the two. It had a shell very similar to the American M-1, but it was fitted with a German style M-53 liner. It was obvious that the American steel helmet played a heavy role in the design of the German-American headgear. They appeared much the same, except that the bottom did not flare out as much as the American

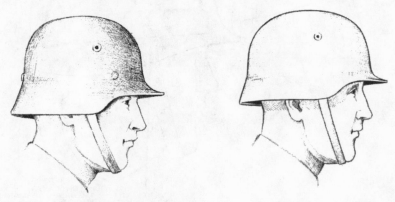

(Left) The early Bundeswehr helmet was basically the same as the old M-35, with the same location of the liner rivets. *(Right)* Another type had a top-suspension liner. Note that there were no rivets on the sides and back.

Inside a German-American Bundeswehr M-62 helmet, showing the top-suspension liner. This liner system was also used in the German style.

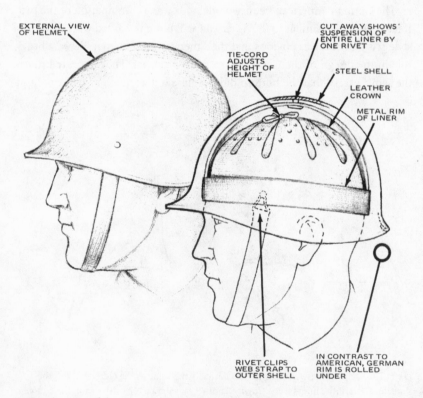

An interior view of the German-American M-62 Stahlhelm.

and the edge was rolled inward smoothly while the American had a separate edging. The M-53 liner system was unique. Only one bolt was responsible for the complete suspension of the liner. It was welded on the inside top of the dome (although it had penetrated through on earlier models) and attached to an inner plastic cup with five plastic, springy bands. These held the thin aluminum liner band. The liner itself was the same as the German M-31. The strap was two-piece webbing that linked together by a spring type clasp.

The other helmet used was a combination of the previous two. It had the German M-35/40 style shell, but with a new type of leather liner assembly suspended by the one-bolt system, much like the M-53 liner. This third helmet was used by the state police. Some had vents and others did not.

There were no insignias on West German helmets except for those worn by certain state police forces. Colors varied but all were based on some shade of green or olive drab.

For many years, the GSG-9 antiterrorist unit of the West German frontier police used a slightly modified fallshirmjäger hemet. It was being phased out in the late 1990s.

In the 1990s, a reunified Germany adopted a new helmet made up of steel leaves covered by ballistic plastic fiber. It resembles the U.S. PASGT but with smaller ear bulges. It is very reminiscent of the old Stahlhelm. The suspension is a unique design.

THE GERMAN STAHLHELM ELSEWHERE

The German Stahlhelm identifies the generic German soldier. At a glance it easily and unmistakably exclaims "Germany!" Thus, the shape of its features is a lasting symbol. It seems to represent that country along with its twentieth-century military history. However, there were others who wore it, and more than a few countries copied its lines.

Since the first viable design, Germany had manufactured and sold its helmets to various governments. Also, the helmet was actually manufactured in other countries with the approval of the Germans—sometimes by license, sometimes without. In those foreign-used pieces there were some small differences: color, badges, or insignias to distinguish proper national and unit identity, liner and strap variations, but they were essentially the Stahlhelm. Helmets captured after both wars were either immediately put to use or later dug out of storage.

In the last decade of the twentieth century, many other helmets have been developed that show Germanic lines. After all, the design is indeed a very effective one. Some of the largest armies today are wearing helmets that were developed along Stahlhelm schematics. The utility of the shape has overcome the brutal image left by two world wars.

GERMAN M-16-17/18	GERMAN M-35	GERMAN FORM
Afghanistan	China	Argentina
Austria	Croatia	Austria
Bulgaria	Germany	Bulgaria
Czechoslovakia	Spain	Chile
Finland	Norway	Czechoslovakia
Germany	Finland	Dominican Republic
Hungary		Eire (Ireland)
Mexico		Finland
Poland		Hungary
Turkey		Spain
Latvia		Turkey
Lithuania		Yugoslavia
USSR		

German Helmets

M-16 WWI helmet. Field Green. Side vent lug. Note: Bottom rivet is location for strap connection lug.

M-17/18 WWI helmet. Field green. Side frontal plate lug. Strap connected to liner band and omits strap lug. Note: Absence of rivet.

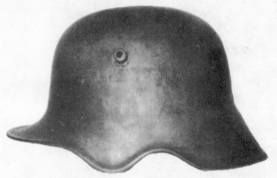

m-18 special cutout WWI helmet. Field green. Side vent lug.
Cutout in ear guard. m-17/18 liner and strap.

WWI camouflage pattern on an m-16 German helmet.

German frontal plate. Field green. Shown on
German helmet with a replacement strap.

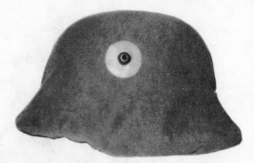

WWI helmet cover. Grey cloth with leather reinforcement around the lughole. String for tying in back.

Frei Korps helmet. A leftover M-17/18 German helmet with a large white painted swastika in upright position.

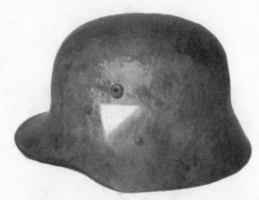

Reichwehr helmet, Bayern. Leftover Austrian model with a white and blue painted Bavarian shield on the left side.

Austrian WWI helmet. Field brown color. Side frontal plate lug. Bottom rivet shows position of strap connection.

Austrian Berndorfer м-17 WWI. Field brown color. Front view shows crimp in front visor.

Austrian Berndorfer WWI (side view). Note the absence of side frontal plate lugs. This connection is located on the extreme top, with breather holes.

Berndorfer frontal plate. Field brown color.

WWI Turkish helmet. Visorless, grey-green color.

м-18 Turkish helmet. Rolled edge.

Hungary, late WWI. Green. Note the rivet location behind the lug.

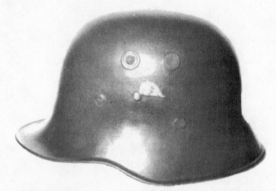

Austrian border guard. Postwar green helmet. WWI style but has screen vents instead of lugs. Aluminum badge on left side.

German SA helmet. Red and gold decal on the right side of a former Austrian Berndorfer m-17 helmet.

Left side of SA Berndorfer helmet. Red and gold eagle on field-brown helmet.

SA Luftschutz three-piece м-38. White insignia on the left side of black beaded Luftschutz helmet.

Right side of the SA Luftschutz helmet. White SA eagle. Four rivets on liner.

Special cutout transitional. All green with Wehrmacht decal on the left and state colors on the right. M-31 liner and strap.

M-42 Wehrmacht helmet. Dark grey with a silver army decal on the left side only. Raw edge.

Transitional from M-16. Green shell with a Wehrmacht decal on the left and the national colors on the right. M-31 liner and strap.

Early Wehrmacht paratrooper. Grey-green with a single decal on the left side.

Officer's aluminum parade. Double decal dress helmet. Very lightweight. Green, smooth, deluxe liner.

Vulkan fiber officer's dress test helmet. Olive green early fiberglass. Wehrmacht decal on the left and the national colors on the right. Large screen-vent. Purpose for oblong hole on earguard was probably for a carrying strap. Extremely lightweight.

M-35 *Wehrmacht helmet.* Green-grey shell. Has the Wehrmacht decal on the left side. M-31 liner and strap. Rolled edge.

Transitional from Austrian M-16. Green shell with a Wehrmacht decal on the left side. M-31 liner and strap.

M-42 *Luftwaffe helmet.* Blue-grey shell. Single Luftwaffe decal on the left side only. Unrolled, raw edge.

Transitional Luftwaffe. Blue-grey shell. M-16 model. Very early Luftwaffe decal on the left side and the national colors on the right.

Early model Luftwaffe. Blue-grey shell. M-35 model. Very early Luftwaffe decal (thicker eagle). Rolled edge.

M-35 *Luftwaffe helmet.* Blue-grey shell. Single decal of the later Luftwaffe type on the left.

Fire service. Black M-35 with a silver stripe and a decal on the left. Aluminum comb on the top.

Jet helmet. Steel layers covered with dark brown leather. Padded comb in the front for shock protection. It was designated SSK90, 1941.

Paratrooper helmet. Grey-green with a single decal on the left side. Rolled edge.

Chicken wire cover for paratrooper. Clamped on the bottom with two clasps.

Afrika Korps helmet. Rolled edge M-35, sand finish of a rough texture, tan color, field applied.

Afrika Korps helmet. M-35 helmet, brown in color with painted insignia on the left.

Russian Front camouflage. Heavy textured white for snow on M-35 model, field applied.

White Kriegsmarine. M-35 with a single gold decal on the left side.

Paratrooper camouflage. Cover with foliage fittings.

Chicken wire cover. Over M-35. Steel wire is octogon pattern of 1½" openings. Fastens to the bottom of the helmet.

Netting cover. Placed over the M-35. Used for attaching foliage.

Nazi camouflage helmet. M-42 rough textured. Has colors of greens and browns in splash pattern, field applied.

Kreigsmarine. Blue-grey M-35 with a gold decal on the left.

Glider helmet. Used in the early 1930s. Made of aluminum.

Variant helmet. WWI design wih a square dip.

104

Police variant. Blue with a short visor and earguard.

Red cross helmet. Luftschutz style body with a single decal on the left side.

Bahnschutz (Railway Police). Very dark green finish with a brass and black metal badge, on an M-35 helmet.

TeNo technical emergency forces model. Unusual shape. Lighter weight. Grey-green color.

Luftschutz Police. Double police insignia on a combat-style beaded Luftschutz.

Combat-style Luftschutz. Dark blue with a silver decal in the front. Body has bead around the base of the dome.

Late model Luftschutz. Dark blue on a one-piece shell.
Four rivets on the liner. Single piece stamping.

Two-piece mid-war model. With bead around crown. Dark
blue with silver insignia in the front.

Three-piece early model. Same as the two-piece with the
exception of having been made with a two-piece visor.

Waffen SS transitional. Apple green color. M-18 shell. SS decal
on the right and the state colors on the left.

Early Allgemeine SS. All-black Austrian M-16
helmet with a transitional liner. SS decal on the
right and the state colors on the left.

Waffen SS helmet. Dark green color. M-35 model.
SS decal on the right side and the state colors on the left.

SS splotch camouflage cover. With foliage fittings.

Early model M-34 Fire/Police. All black with a large
slanted white swastika on the right and slanted national
colors on the left. Leather neck protector. Cup-type
vents.

Police variant. All black with a large police decal on
the left. Lightweight. Square dip on the ear guard.

Army Field Police helmet. Heavy M-35 model. All field green with a police decal on the left side.

Right side of Field Police helmet. Has Nazi party decal. Regular M-31–type liner

Wasserschutzpolizei (Water Police). M-31 liner. Field green, heavy M-35 model. Gold police decal on the left. State colors on the right.

110

Lightweight police. Police-type liner. Square dip on ear guard. Silver decal on the left, state colors on the right. All black.

Police/Fireman's helmet. Lightweight. All black with a silver police decal on the left and the state colors on the right. Leather neck protector. Square dip. Aluminum comb.

Early model police/fireman. Black WWI-profile helmet with no lugs. Police decal on the left side. Aluminum comb.

Czech м-1932 re-issue helmet. Dark blue
with Luftschutz insignia in the front.

French м-1926 re-issue helmet. Black with the Wehrmacht
and national colors on the reversed sides.

Czech м-1932 re-issue helmet. Brown with the
police insignia on the left side.

Danish M-1923 re-issue helmet. Black with Luftschutz decal in the center. Refitted liner.

Austrian postwar German M-16. Marine Ehrhardt Brigade brown with a metal badge.

Austrian paratrooper. American design but has German-style liner. Brown-grey with rolled edge.

Hungarian WWII helmet. Similar to German. Note the rivet locations and attachment in the back for hanging.

West German paratrooper and GSG-9. Very similar to WWII model but had a revised harness system.

West German M-62. German-American–style shell with an M-31 liner suspended from a bolt on the top.

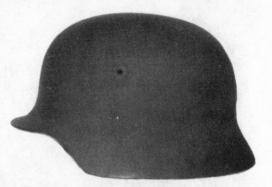

West German. German WWII style but with a new suspension liner. Note: no rivets showing.

East German helmet. Green-grey color. Left view.

East German helmet. Front view. Note position of rivets.

East German paratrooper. The small number of DDR airborne troops were equipped with Polish model 1963 light steel helmets, as were a number of other Warsaw Pact forces.

East German border police. This postwar-designed squared-off Germanic style helmet was allowed by the USSR for use by the border troops until the 1960s when it, like all other variants, was replaced by the ubiquitous M-56.

Para-military auxiliaries. Soviet-made helmet used by the workers' militia of the initial Soviet-run East German workers and student guards, thought to be more ideologically reliable than the rest of East Germany's population. Eventually these old Soviet M-40 models were replaced by high-quality Czech-made M-53s. The Czech helmets were used into the early 1960s.

Current United Germany. Along with many other nations, Germany in the 1990s began experimenting with kevlar-like fiber plastics. Germany has now adopted this polyethylene helmet with steel leaves incorporated. This model has also been adopted by other NATO nations including the Netherlands and Denmark.